Learning Materials in Biosciences

Learning Materials in Biosciences textbooks compactly and concisely discuss a specific biological, bio-medical, biochemical, bioengineering or cell biologic topic. The textbooks in this series are based on lectures for upper-level undergraduates, master's and graduate students, presented and written by authoritative figures in the field at leading universities around the globe.

The titles are organized to guide the reader to a deeper understanding of the concepts covered.

Each textbook provides readers with fundamental insights into the subject and prepares them to inde-pendently pursue further thinking and research on the topic. Colored figures, step-by-step protocols and take-home messages offer an accessible approach to learning and understanding.

In addition to being designed to benefit students, Learning Materials textbooks represent a valuable tool for lecturers and teachers, helping them to prepare their own respective coursework.

More information about this series at http://www.springer.com/series/15430

Tuck Seng Wong • Kang Lan Tee

A Practical Guide to Protein Engineering

Springer

Tuck Seng Wong
Department of Chemical & Biological
Engineering
University of Sheffield
Sheffield, UK

Kang Lan Tee
Department of Chemical & Biological
Engineering
University of Sheffield
Sheffield, UK

ISSN 2509-6125   ISSN 2509-6133   (electronic)
Learning Materials in Biosciences
ISBN 978-3-030-56897-9   ISBN 978-3-030-56898-6   (eBook)
https://doi.org/10.1007/978-3-030-56898-6

This Springer imprint is published by the registered company Springer Nature Switzerland AG
The registered company address is: Gewerbestrasse 11, 6330 Cham, Switzerland

This book is dedicated to our parents in Kuala Lumpur, Singapore and Sheffield, whom we love dearly.

Preface

Motivation

Protein engineering is our research passion and a subject that is close to our hearts. We have always wanted to write a textbook to summarise and share our experience in protein engineering, but never got down to it until now. Although there are excellent edited books on the topic of protein engineering, they are mostly targeted at research scientists or more advanced protein engineers. To our best knowledge, there isn't an authored book on protein engineering. There is also no protein engineering textbook written specially for university students. When this opportunity from Springer Nature arose, we accepted the challenge to fill this gap.

Target Audience and Prerequisites

This book is targeted at advanced undergraduate students in their 3rd or 4th year of study, postgraduate students and young researchers wanting to acquire protein engineering knowledge. We have written this book based on the assumption that you have a basic knowledge of biochemistry and molecular biology and you are familiar with the scientific terms listed below:
- Biochemistry (DNA, RNA, protein, codon, gene, operon, transcription, translation, enzyme, enzyme inhibition)
- Molecular biology (cloning, plasmid, polymerase chain reaction, restrictive enzyme, agarose gel electrophoresis, restriction digestion, ligation, transformation, DNA sequencing)

Don't despair, if you have not encountered these terms. We recommend that you first build your foundation by referring to the textbooks below, before reading this book. These are excellent textbooks that we highly recommend to our students:
- Biochemistry, by Donald Voet and Judith G. Voet, John Wiley & Sons
- Gene Cloning & DNA Analysis: An Introduction, by Terry A. Brown, Wiley-Blackwell
- Introduction to Protein Structure, by Carl Branden and John Tooze, Routledge
- Structure and Mechanism in Protein Science: A Guide to Enzyme Catalysis and Protein Folding, by Alan R. Fersht, Kaissa Publications

Writing Styles

We both were once students like you. We fully understand the 'picture superiority effect' and the 'text-organization effect'. Visuals captivate and they help to explain complex concepts. Effective visuals improve learning. Students often have a better memory for pictures than for corresponding words. Students' comprehension is also influenced by the text structure used to convey the information. Our brains like organized information. That explains the large number of figures and tables you'll find in this book.

Whenever possible, we have written in point form to keep the content succinct and easy to digest. Having taught in the higher education sector for over 10 years, we understand that students find long-winded text tiring. Another point to note, we have used personal pronouns throughout this book, which is unorthodox. The intention is to better engage with our learners by communicating directly with you.

The Case Study

Perhaps influenced by our original education and training in engineering, we firmly believe in 'learning through practice'. That's the reason why we have created Case Study 1 on protein engineering of cellulolytic enzymes. We have used this example throughout the entire book to illustrate how to engineer a protein. By doing so, we also hope to achieve cohesion and coherence in the learning material.

The Organization

This book is organised in a way that reflects precisely how we usually approach a protein engineering project. Each chapter begins with the learning objective, describing what we expect a student to be able to understand and apply at the completion of the chapter involved. Further, we have included an exercise at the end of each chapter. These exercises are intended to:
- Reinforce your learning
- Self-assess your understanding
- Encourage your active learning through exploring more advanced topics in the recommended further readings

The Companion Website

We have also created a companion website (▶ https://sites.google.com/sheffield. ac.uk/proteinengineering/) for this book, where you'll find:
- Supplementary materials for each chapter to facilitate your learning (*e.g.*, gene sequences and protein structures we use in this book)
- Quizzes for each chapter for you to assess your own learning

Feedback

The bulk of this book was written during the COVID-19 lockdown in the UK. We were both registered as key workers in the University of Sheffield, as we volunteered to manufacture SARS-CoV-2 antigen and antibody for the Northern General Hospital and the Royal Hallamshire Hospital in Sheffield. Despite spending long hours in the laboratory, we dedicated every spare minute we had on completing this book. Juggling between lab work, book writing and other academic commitments has been extremely challenging. If you spot mistakes that we overlooked, please accept our apology and kindly let us know. We welcome your feedback and suggestion.

Tuck Seng Wong
Sheffield, UK

Kang Lan Tee
Sheffield, UK

Acknowledgement

We would like to start by thanking our parents, sisters and brothers for supporting our academic and research careers in the United Kingdom.

We are thankful for all the opportunities given throughout our career to learn from some of the finest scientists in the world: Nobel Laureate Prof. Frances H. Arnold, Prof. Sir Alan R. Fersht, Prof. Ulrich Schwaneberg, Prof. Andrew W. Munro and Prof. Alexander Steinbüchel. A heartfelt thank you for your guidance and support.

We thank our former colleague Prof. Miguel Alcalde for his kind recommendation to Springer Nature, without which this book project would not have been possible.

Finally, to all our colleagues at the University of Sheffield who have supported us all these years: Prof. Phillip Wright, Prof. Catherine Biggs, Prof. Ray Allen and Prof. James Litster.

Contents

About the Authors

Tuck Seng Wong

is a Senior Lecturer (Associate Professor) in the Department of Chemical and Biological Engineering at the University of Sheffield (UK). He is also a Visiting Professor at the National Center for Genetic Engineering and Biotechnology (BIOTEC) in Thailand. Dr. Wong leads a Biocatalysis and Synthetic Biology group in Sheffield, and his research focuses on sustainable biomanufacturing using engineered enzymes or microbes. His passion in engineering of biology is inspired by his research supervisors and mentors including Nobel Laureate Prof. Frances H. Arnold (Caltech), Prof. Ulrich Schwaneberg (Bremen), Prof. Sir Alan R. Fersht (Cambridge) and Prof. Alexander Steinbüchel (Münster). Dr. Wong obtained his BEng in Chemical Engineering (1st class Honours) from the National University of Singapore, followed by an MSc and a PhD (special distinction, highest possible grade) in Biochemical Engineering from the Jacobs University Bremen in Germany. He is a recipient of multiple prestigious fellowships, including the RAEng|The Leverhulme Trust Senior Research Fellowship (2019), the Royal Academy of Engineering Industrial Fellowship (2016) and the MRC Career Development Fellowship (2007–2009). He was one of the 10 young academics appointed to the EPSRC Early Career Forum in Manufacturing Research in 2014 and a world finalist of the Synthetic Biology Leadership Excellence Accelerator Program (LEAP) in 2015. He has published over 45 papers, mostly in the areas of protein engineering and synthetic biology.

Kang Lan Tee

is a Lecturer (Assistant Professor) in the Department of Chemical and Biological Engineering (CBE) at the University of Sheffield (UK). She has published over 20 papers in the field of protein engineering and synthetic biology since 2004. Dr. Tee is also the co-Founder of the protein engineering start-up SeSaM-Biotech GmbH and a Global Challenges Research Fellow (2019–2021). Her research team applies protein engineering strategies to address global challenges in biomanufacturing and sustainability. She also develops and leads the biorefineries module in CBE.

Literature Search

Contents

© Springer Nature Switzerland AG 2020
T. S. Wong, K. L. Tee, *A Practical Guide to Protein Engineering*, Learning Materials in Biosciences,
https://doi.org/10.1007/978-3-030-56898-6_1

1

What You Will Learn in This Chapter
In this chapter, we will learn to:
- identify and use academic search engines
- perform a simple and an advanced search in PubMed
- filter search results
- extract key information from a literature search
- understand the structure of an original article and extract information from relevant sections

Perhaps influenced by our original education and training in engineering, we firmly believe in 'learning through practice'. That's the reason why we have created Case Study 1 on protein engineering of cellulolytic enzymes. We will use this example throughout the entire book to illustrate how to engineer a protein.

Case Study 1

Your project supervisor has come across a soil cellulolytic bacterium *Thermobifida fusca*, and would like you to prepare an enzyme cocktail for cellulose degradation using enzymes originally derived from this bacterium. If necessary, you will use protein engineering approach to enhance the enzyme performance.

A research project, like Case Study 1, is best begun with an extensive literature search. In fact, scientific literatures often stimulate novel ideas and polish existing ones, both of which are critical for executing a successful project. This easily and often neglected aspect is as important as collecting experimental data in the lab for various reasons:
- To understand the significance of the project (*e.g.*, What are the potential research impacts of your project? Who are the beneficiaries? How does it benefit the society? How does it solve global challenges?)
- To build background knowledge on the research topic (*e.g.*, What have been done in this area? Who are the leading scientists in the field? Where does your project fit in? How would your project contribute to the field? Where does the novelty lie?)
- To expand technical knowledge and experimental skills (*e.g.*, What are the commonly used experimental approaches? Are there useful experimental techniques that can be adopted? What are the potential technical challenges?)
- To hone technical writing skills and use of scientific terminology (*e.g.*, What is the format of a technical article? How to write better?)

1.1 Databases

Conducting a thorough literature search and keeping abreast of newly published papers are challenging, but highly rewarding and satisfying tasks. How do we find and access scientific literatures? There are many different academic search engines, and they differ substantially in terms of coverage and retrieval qualities. Some of these search engines are free, but some require subscription. Some focus on a single

◘ **Table 1.1**   Academic search engines relevant to protein engineering

Academic search engine	Uniform resource locator (URL)	Free/ Subscription	Discipline
PubMed	► https://pubmed.ncbi.nlm.nih.gov/	Free	Life Sciences
Google Scholar	► https://scholar.google.com/	Free	All
Web of Science	► https://clarivate.com/webofsciencegroup/ solutions/web-of-science/	Subscription	All
Scopus	► https://www.scopus.com/	Subscription	All

discipline, while others cover multiple fields. In ◘ Table 1.1, we have summarized our favourite search engines for protein engineering research.

Of all the search engines, PubMed is the one that we most often use. Made publicly available since 1996, PubMed was developed and is maintained by the National Center for Biotechnology Information (NCBI). It is a free search engine, accessing primarily the MEDLINE database of references and abstracts on life sciences and biomedical topics. Most protein engineering literatures can therefore be retrieved using PubMed. For student researchers, we would highly recommend that you kickstart your academic literature search using PubMed.

1.2 How to Search?

Conducting a literature search is identical to doing a Google search, a familiar task for many of us. The key steps are:

1. Identify the keywords of your project
2. Enter the webpage of the search engine (*i.e.*, use the URL provided in the ◘ Table 1.1 that corresponds to the search engine you want to use)
3. Type these keywords individually or in combination in the search field or query box
4. Click search
5. Result will be displayed in a summary format

The search result is dependent on the keyword or the keyword combination applied. If we take Case Study 1 in this book as an example, potential keywords to be used in our literature search include `soil cellulolytic bacterium`, `Thermobifida fusca`, `enzyme cocktail`, `cellulose degradation` and `protein engineering` (◘ Fig. 1.1). If you search with a single general keyword say `cellulose degradation`, PubMed will return 47680 results (as of 20th April 2020). It is practically unfeasible to read or scan through these vast number of literatures. You can narrow your search down by using two keywords in combination such as `cellulose degradation` and `Thermobifida fusca`. With this combination, PubMed returns a total of 123 results (as of 20th April 2020), which is far more manageable.

1

□ **Fig. 1.1** Potential keywords (highlighted in yellow) in Case Study 1 that can be applied in a literature search.

Your project supervisor has come across a soil cellulolytic bacterium *Thermobifida fusca*, and would like you to prepare an enzyme cocktail for cellulose degradation using enzymes originally derived from this bacterium. If necessary, you will use protein engineering approach to enhance the enzyme performance.

If we are new to a research topic, it is often helpful to first read a couple of comprehensive review articles. Review articles provide an extensive summary of the research on a certain topic, a perspective on the state of the field, where it is heading, and future prospects. It is the fastest way to get acquainted with the topic. There are two approaches to retrieve review articles:

1. You can do a quick literature search using a keyword in combination with the word `review`. If we conduct a PubMed search using `Thermobifida fusca` and `review`, only three articles are returned (as of 20th April 2020). One of these three articles is a review article published in Critical Reviews in Microbiology on the topic of '*The cellulolytic system of Thermobifida fusca*',[1] which is precisely what we are looking for.
2. Alternatively, you can use the filter function located on the left sidebar of PubMed. If we conduct a PubMed search using `Thermobifida fusca`, we will get a list of 303 articles (as of 20th April 2020). If we tick the 'Review' box under 'Article Type', we are left with three articles (as of 20th April 2020) that are identical to those obtained using the first approach.

As you become more experienced with the PubMed search, you may want to try the advanced search feature to further refine your search. To do so, you simply click on the word 'Advanced' located beneath the query box in ► https://pubmed.ncbi.nlm. nih.gov/ or go directly to ► https://pubmed.ncbi.nlm.nih.gov/advanced/. Advanced search allows you to more precisely define your search criteria. Using the same example described above, you want to collect review articles on *T. fusca*. All you need to do is to add the following two search terms into the query box, click search, and you will get the same list of three review articles (as of 20th April 2020):

- `Thermobifida fusca[Title/Abstract]`
- `Review[Publication Type]`

The first search term `Thermobifida fusca[Title/Abstract]` means the word *Thermobifida fusca* must appear either in the title or in the abstract. The second search term `Review[Publication Type]` specifies the article type as review.

You may have noticed by now that we keep on emphasizing 'as of 20th April 2020'. This is because literature volume is expanding on a daily basis. The search results will therefore vary, depending on when the literature search is conducted. By the time you read this book, the number might have doubled, tripled, quadrupled, or more.

---

1   Gomez del Pulgar, E.M. et al., (2014). The cellulolytic system of *Thermobifida fusca*. *Critical Reviews in Microbiology*. **40**(3), 236–47. Available from DOI: ► https://doi.org/10.3109/10408 41X.2013.776512

1.3 Important Information

Once you have identified an article of interest, the next step is to extract useful information from it. Let's use the review article *'The cellulolytic system of Thermobifida fusca'* published in Critical Reviews in Microbiology (■ Fig. 1.2) as an example to illustrate the types of information you can get (■ Table 1.2).

Report writing is a common task for students, and referencing is an essential aspect of a convincing report. Information indicated by an asterisk (*) in ■ Table 1.2 are commonly used to cite an article. Other information like DOI and PMID are also useful for retrieving articles using referencing software like EndNote. Interesting to note, mistakes in citations are most commonly seen in the author's name. A name typically comprises first name (given name), middle name, and last name (family name or surname). In ■ Table 1.3, we have summarized some of the common mistakes we have seen in reporting an author's name.

After reading a couple of reviews, one would progress to original research articles to understand specific pieces of research work and their research methodologies. Let's use an article *'Quantitative iTRAQ secretome analysis of cellulolytic Thermobi-*

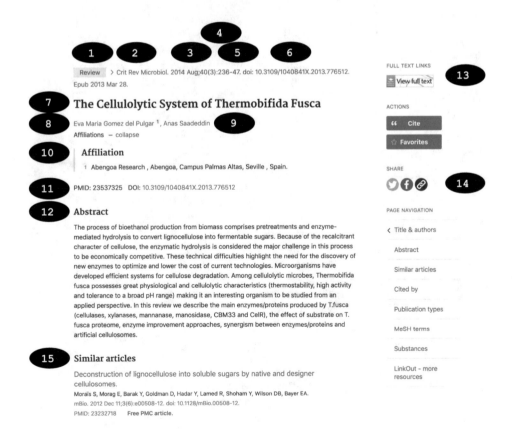

■ **Fig. 1.2** Key information obtained from a literature search.

1

□ Table 1.2   Key information obtained from a literature search

Number	Information type	Specific to the example used	How is the information useful to you?
1	Article type	Review	It shows you the type of article you will be reading.
2*	Journal	Critical Reviews in Microbiology, which is abbreviated as Crit Rev Microbiol	This is the scientific journal where this article was originally published.
3	Publication date	Aug 2014	This tells you when the article was published and suggests how updated the information is relative to current state-of-the-art.
4*	Volume (Issue)	Volume 40, issue 3	These are the volume and the issue numbers of the journal where this article was published.
5*	Page number	236 till 247, often written in number range of 236–47 or 236–247	These are the pages in the issue where this article was published.
6	Digital object identifier (DOI)	▶ https://doi.org/10.3109/1040841X.2013.776512	This unique alphanumeric string is assigned to an online document such as a journal article or an e-book. A DOI can be used in a similar way to a URL, but it is more permanent and reliable. DOI is usually written in two formats: ▬ DOI: xxx (*e.g.*, DOI: 10.3109/1040841X.2013.776512) ▬ https://doi.org/xxx (*e.g.*, ▶ https://doi.org/10.3109/10408 41X.2013.776512)
7*	Title	The Cellulolytic System of *Thermobifida fusca*	This is the title of the article.
8*	First author	Eva Maria Gomez del Pulgar, which is abbreviated as Gomez del Pulgar EM	If you enjoy following the research work from this author or the articles written by this author, you can search for other articles from the same author, *e.g.*, using the search term Gomez del Pulgar EM[Author].
9*	Last author	Anas Saadeddin, which is abbreviated as Saadeddin A	Same as above.

◻ **Table 1.2** (continued)

Number	Information type	Specific to the example used	How is the information useful to you?
10	Affiliation	Abengoa Research, Abengoa, Campus Palmas Altas, Seville, Spain	The work address of these authors is useful if you wish to contact them.
11	PubMed identifier (PMID)	23537325	This unique identifier number retrieves this article from the PubMed database.
12	Abstract	The process of bioethanol ...... artificial cellulosomes.	The abstract provides a summary of the content and key findings presented in the article. It is useful for assessing its suitability and quality beyond the broad information derived from the title.
13	Full text links	Taylor & Francis icon with the word 'View full text' embedded	For open access articles, you can read the full text or download the Portable Document Format (PDF) of this article. Some journals require subscription. If your academic or research institution subscribes to these journals, you can also read or download articles from these journals.
14	Sharing	Twitter icon, Facebook icon, and Permalink icon	You can promote your favourite articles via social media or promote your own articles to generate impact and awareness for your work.
15	Similar articles	List of articles beneath the abstract	This list allows you to quickly access related articles.

*Common information used to cite an article.

fida fusca[2] published in the Journal of Proteome Research to explain the structure of an original research article and the useful information you can expect. Most original articles, if not all, are divided into clearly defined sections that include 'Introduction', 'Materials & Methods', 'Results', 'Discussion', 'Conclusion', 'Acknowledgement', 'Supporting Information Available', and 'References'. Under the 'Materials & Methods', 'Results', and 'Discussion' sections, you often find subsections with their corresponding subheadings as well. This organization allows readers to quickly find the information they need, as illustrated in ◻ Table 1.4.

---

2   Adav, S.S. et al., (2010). Quantitative iTRAQ secretome analysis of cellulolytic *Thermobifida fusca*. *Journal of Proteome Research*. **9**(6), 3016–24. Available from DOI: ▶ https://doi.org/10.1021/pr901174z

1

■ Table 1.3 Common mistakes in reporting an author's name

Common mistake	Wrong ✘	Correct ✓	Explanation
Umlaut	Johannes Muller	Johannes Müller	As well as the 26 letters of the alphabet, the German language is also characterised by the umlaut, a diacritic in the form of two dots placed over the letters 'a', 'o', and 'u' to form 'ä', 'ö', and 'ü'.
Preposition	Vincent van Gogh is abbreviated as 'Gogh VV'	van Gogh V	There are surnames that use a particle, prefix, or preposition, such as 'van' (written in small letters).
Double-barrelled surnames	Andrew Lloyd Webber is abbreviated as 'Webber AL'	Lloyd Webber A	A name may carry two surnames from two families.
Arabic name	Mahathir bin Mohamad is abbreviated as 'Mohamad MB'	bin Mohamad M	In Arabic names, both 'ibn' and 'bin' is translated as 'son of'.
Chinese names in 3 words	Lee Kuan Yew is abbreviated as 'LK Yew'	KY Lee	When Chinese names are written in Chinese characters (李光耀), the last name (李) always appears first. Some Chinese names have three words (*e.g.*, 'Lee Kuan Yew'). In this example, 'Kuan Yew' is the first name and 'Lee' is the last name.

■ Table 1.4 The structure of an original article and the information from each section

Section	Type of information that can be found	Examples
Introduction	• Background of the research • Objectives of the study • Unique contribution of this study	• Background on lignocellulosic biomass degradation • The objective of this study was to identify and quantify the predominant secretary proteins involved in cellulose and lignin hydrolysis
Materials & Methods	• Details of the work that allow the experiments to be reproduced by an independent researcher	• Chemicals used, their suppliers, and product numbers • Media/buffer composition • Cultivation conditions of *T. fusca* • Secretome extraction procedures • Proteomics procedure and data analysis • Enzymatic assay for cellulose hydrolysis

◘ Table 1.4 (continued)

Section	Type of information that can be found	Examples
Results	• Experimental data typically presented in the format of figures (*e.g.*, graphs, pictures, *etc*) and tables	• Proteins secreted by *T. fusca* (*e.g.*, cellulases, hemicellulases, other glycoside hydrolases, proteolytic enzymes, *etc*) under different culture conditions (*e.g.*, in the presence of cellulose, lignin, and cellulose + lignin)
Discussion	• The significance of the result. This section is sometimes combined with the 'Result' section	• Discussion on various protein classes secreted by *T. fusca* • Molecular mechanism of protein secretion by *T. fusca* • Biotechnological applications of enzymes secreted by *T. fusca*
Conclusion	• Conclusion of the study • Some authors might briefly describe their future work	• This study reported the comprehensive systematic quantification and regulation of *T. fusca* secretome constituent including cellulases, hemicellulases, other glycoside hydrolases, proteases, and others
Acknowledgement	• A list of individuals who provided help during the research (*e.g.*, providing samples, providing language help, writing assistance or proofreading the article, *etc*) • A list of organizations who provided financial support (*e.g.*, funding agencies, scholarship providers, *etc*)	• Funding agencies and grant codes
Supporting Information Available	• Additional information on Materials & Methods • Additional figures and tables • DNA or protein sequences • Raw data	• An additional table on proteomics data is available in the supplementary material
References	• A list of references cited in the article	• A total of 37 references were cited in the article

---

┌─ **Take-Home Messages**

1. A literature search is a fundamentally important aspect of executing a successful research project.
2. A literature search helps the researchers to understand the significance of the research project, build background knowledge on the research topic, expand technical knowledge and experimental skills, and hone technical writing.
3. PubMed is one of the most commonly used academic search engines in the field of protein engineering.

1

4. The retrieval quality of a literature search is determined by the academic search engine used and the keywords applied in the search.
5. Mistake in author's name is a common problem in referencing and science communication.
6. An abstract is a vital part of any research article. It should provide a succinct summary of the research study, and encourage the readers to read further.
7. Most scientific articles share a common format. They are divided into distinct sections and each section contains a specific type of information.
8. Scientific articles comprise the following parts: Introduction, Materials & Methods, Results, Discussion, Conclusion, Acknowledgement, Supporting Information Available and References.

Exercise

Case Study 2

Scientists from the Kyoto Institute of Technology isolated a bacterium called *Ideonella sakaiensis* 201-F6 at a poly(ethylene terephthalate) (PET) bottle recycling facility in Sakai, Japan. This bacterium uses the PET plastic as its primary carbon and energy source (Science 2016, DOI: ▶ https://doi.org/10.1126/science.aad6359). There are two enzymes in this bacterium that are responsible for breaking PET down into its monomers, *i.e.*, terephthalic acid and ethylene glycol. Your supervisor would like you to engineer a polyesterase for enhanced plastic degradation.

(a) List the potential keywords that can be used in a PubMed search to understand enzymatic plastic degradation.
(b) What is the title of this article that appeared in Science in year 2016?
(c) Who is the first author of this piece of work?
(d) What are the two enzymes responsible for converting PET into its monomers, terephthalic acid and ethylene glycol?

Further Reading

Ecker ED, Skelly AC (2010) Conducting a winning literature search. Evid Based Spine Care J 1(1):9–14
Falagas ME, Pitsouni EI, Malietzis GA, Pappas G (2008) Comparison of PubMed, Scopus, Web of Science, and Google Scholar: strengths and weaknesses. FASEB J 22(2):338–342
Fiorini N, Canese K, Starchenko G, Kireev E, Kim W, Miller V, Osipov M, Kholodov M, Ismagilov R, Mohan S, Ostell J, Lu Z (2018) Best match: new relevance search for PubMed. PLoS Biol 16(8):e2005343
Leonard SA, Littlejohn TG, Baxevanis AD (2007) Common file formats. Curr Protoc Bioinformatics, Appendix 1, Appendix 1B

Sequence Analysis

Contents

© Springer Nature Switzerland AG 2020
T. S. Wong, K. L. Tee, *A Practical Guide to Protein Engineering*, Learning Materials in Biosciences,
https://doi.org/10.1007/978-3-030-56898-6_2

2

What You Will Learn in This Chapter

In this chapter, we will learn to:

- describe the chemical structure of cellulose
- specify the key enzymes involved in cellulose degradation
- search for a gene and obtain its nucleotide and protein sequences
- create and save a sequence in its FASTA format
- analyse a protein sequence
- analyse a DNA sequence
- use bioinformatics tools

Recall the task in Case Study 1, where you are asked to create an enzyme cocktail for cellulose degradation using enzymes from *T. fusca*. By now, you should have conducted your literature search and acquired some background knowledge on cellulose degradation by *T. fusca*. Cellulose is an essential structural component of plant cell wall and is the most abundant organic polymer on earth. It is a linear polysaccharide chain of repeating D-glucose (or β-D-glucopyranose) units connected by β(1→4) glycosidic bonds. From literature search, you would have identified the key enzymes required for a complete cellulose hydrolysis to glucose, and these include endo-β-1,4-glucanases, exoglucanases or cellulose 1,4-β-cellobiosidase, and β-1,4-glucosidases (☐ Table 2.1). If you wish to find out more about these enzymes, we would refer you to the Carbohydrate-Active Enzymes Database (CAZy; ► http://www.cazy.org/). The CAZy database describes the families of structurally-related catalytic and carbohydrate-binding modules (or functional domains) of enzymes that degrade, modify, or create glycosidic bonds.

2.1 Gene Information

At this point, you might ask:

- Where can I find these genes that encode the enzymes?
- How many endo-β-1,4-glucanase-encoding genes are there in the *T. fusca* genome?

☐ **Table 2.1**   Key enzymes involved in cellulose degradation

Key enzymes	Enzyme Commission (EC) number	Enzymatic action	Synonyms
Endo-β-1,4-glucanases	3.2.1.4	Hydrolyse the internal glycosidic bonds in cellulose	Endoglucanases (EG) / cellulases (in some literatures)
Exoglucanases	3.2.1.91	Hydrolyse cellulose from the nonreducing or reducing chain ends to release cellobiose	Cellulose 1,4-β-cellobiosidases / cellobiohydrolases (CBH)
β-1,4-Glucosidases	3.2.1.21	Hydrolyse cellobiose to release glucose	β-Glucoside glucohydrolase / β-glucosidases (BGL)

- How many exoglucanase-encoding genes are there in the *T. fusca* genome?
- How many β-1,4-glucosidase-encoding genes are there in the *T. fusca* genome?
- How do I identify these genes?

We will now show you three different methods (KEGG, NCBI, and SnapGene) of searching for a gene. For each of these methods, we will follow three main steps:
1. Get the genome of *T. fusca*
2. Search for genes encoding for an enzyme of interest (*e.g.*, endo-β-1,4-glucanase)
3. Extract the corresponding gene sequence

2.1.1 Kyoto Encyclopedia of Genes and Genomes (KEGG)

KEGG is a database resource for understanding high-level functions and utilities of the biological system, such as the cell, the organism, and the ecosystem, from molecular-level information, especially large-scale molecular datasets generated by genome sequencing and other high-throughput experimental technologies.

To search for endoglucanase-encoding genes in *T. fusca* genome:
1. Get the genome of *T. fusca*
    - Go to the KEGG webpage (▶ https://www.genome.jp/kegg/)
    - In the 'KEGG' query box, type Thermobifida fusca
    - Click the 'Search' button
    - Under 'KEGG Genome' in the result page, click 'T00265' that corresponds to the genome of *T. fusca* YX
2. Search for endoglucanase-encoding genes
    - Click 'Show organism' under the entry 'Annotation'
    - In the 'Search gene' query box, type Endoglucanase
    - Click the 'Go' button
    - A list of five endoglucanase genes (Tfu_0901, Tfu_1074, Tfu_1627, Tfu_2176, and Tfu_2712) will be displayed
3. Extract the corresponding endoglucanase gene sequence
    - To find out more about gene Tfu_0901, click 'Tfu_0901'
    - Under the entry 'AA seq' (refers to amino acid sequence), you will find the protein sequence MAK ...... LQS
    - Under the entry 'NT seq' (refers to nucleotide sequence), you will find the DNA sequence atggcgaaa ...... cagtcctga

KEGG is an excellent database that provides vast amount of information. For students interested in protein engineering, you may find the following information useful for your project:
- Under the entry 'Structure', you will find the protein structures of this endoglucanase with the Protein Data Bank (PDB) codes of 2CKS and 2CKR.
- Under the entry 'Position' and when you click on the 'Genome map' button, you will see where the gene is located within the genome relative to other genes (*e.g.*, Is the gene located within a cellulase gene cluster? Is the gene part of an operon?). Each gene is colour coded. Tfu_0901 is coloured in blue, indicating that this gene is involved in the carbohydrate metabolism.

2

2.1.2 **National Center for Biotechnology Information (NCBI)**

In ▸ Chap. 1, we have briefly introduced NCBI, where PubMed is hosted. NCBI is a great resource for genomics, genetics, and biomedical data.

To search for endoglucanase-encoding genes in *T. fusca* genome:

1. Get the genome of *T. fusca*
   - Go to the NCBI webpage (▸ https://www.ncbi.nlm.nih.gov/)
   - On the left sidebar, click 'Genomes & Maps'
   - Click 'BioProject (formerly Genome Project)'
   - Under 'Browse BioProject', click 'By Project Attributes'
   - In the query box, type `Thermobifida fusca`
   - Click the 'Search' button
   - A total of 148 results will be returned. Click 'Filter' and tick 'Genome sequencing' under the 'Data type'
   - The list of results is shortened to nine entries, with only one entry corresponding to the genome of *T. fusca*
   - Click the accession 'PRJNA94' that corresponds to the genome of *T. fusca* YX
   - Under the 'Project Data', click the link that corresponds to 'Nucleotide (Genomic DNA)'
   - The complete genome of *T. fusca* YX (GenBank accession number: CP000088.1) will be displayed
2. Search for endoglucanase-encoding genes
   - Press ⌘+F to search
   - In the search field, type `Endoglucanase`
   - You will get 3 CDSs (refers to coding sequences) that are annotated as endoglucanases (Tfu_1074, Tfu_1627, and Tfu2176)
   - In the search field, type `Cellulase`
   - You will get 2 additional CDSs that are annotated as cellulases (Tfu_0901 and Tfu_2712)
3. Extract the corresponding endoglucanase gene sequence
   - To find out more about gene Tfu_0901, click the 'CDS' of Tfu_0901 and select 'GenBank'
   - The protein sequence and the DNA sequence of Tfu_0901 will be displayed

You might have noticed that two terms `Endoglucanase` and `Cellulase` are used for the search in NCBI compared to just `Endoglucanase` in KEGG. Different synonyms (see ◘ Table 2.1) are used to annotate enzymes in different databases and more than one enzyme name may be required to identify all the sequences. Although the NCBI method may seem more complex, the database is highly integrated with other useful tools or resources. For example, when you are in the summary page describing gene Tfu_0901, you can perform a sequence analysis immediately by using the tools listed under 'Analyze this sequence' on the right sidebar, such as Run BLAST. We will cover gene analysis tools in greater depth in ▸ Sect. 2.2.

2.1.3 SnapGene

The next method we are introducing requires that you first download a free software called SnapGene Viewer, a software that allows molecular biologists to create, browse, and share richly annotated DNA sequence files up to 1 Gbp in length. You can download the software from ▶ https://www.snapgene.com/snapgene-viewer/, and the software works for most operating systems including Windows, macOS, and Linux.

To search for endoglucanase-encoding genes in *T. fusca* genome:

1. Get the genome of *T. fusca*
   - Go to SnapGene Viewer
   - Click 'Import' and select 'NCBI Sequence'
   - Type the accession number of the *Thermobifida fusca* YX genome, which is CP000088.1 (You will have to first identify the accession number using NCBI)
   - Click the 'Import' button
   - SnapGene Viewer will download the genome sequence from NCBI
   - 'Save' the file at your preferred location
2. Search for endoglucanase-encoding genes
   - Click 'Features' in the bottom toolbar
   - In the 'Find' query box (located just above the bottom toolbar), type Endoglucanase
   - Click the 'Next' button
   - You will see three genes annotated as endoglucanases (Tfu_1074, Tfu_1627, and Tfu2176)
   - If you find using Cellulase, you will see two additional genes annotated as cellulases (Tfu_0901 and Tfu_2712)
3. Extract the corresponding endoglucanase gene sequence
   - To find out more about gene Tfu_0901, click 'Tfu_0901' and the word 'Tfu_0901' should now be highlighted in blue colour
   - Click 'Sequence' in the bottom toolbar
   - The DNA sequence encoding 'Tfu_0901' is shown and highlighted in light blue colour
   - To copy the protein sequence, click 'Edit' in the navigation menu, select 'Copy Amino Acids' and 'Copy 1-Letter Amino Acids'
   - To copy the coding strand, click 'Edit' in the navigation menu, select 'Copy Bottom Strand Bases' and '5'→3' (atggcg…)' [You will need to judge which strand (top/bottom) to copy. In the case of Tfu_0901, the bottom strand is the coding strand.]

If you intend to work frequently with the *T. fusca* YX genome, we would strongly advise that you create a SnapGene file of this genome and search the genome using the steps outlined above.

2.1.4 Other Databases of Protein Sequences

UniProt is a freely accessible database of protein sequence and functional information, with many entries being derived from genome sequencing projects. If you wish to search for a specific protein sequence (*e.g.*, Tfu_0901), you may want to consider using UniProt:

2

- Go to the UniProt webpage (▶ https://www.uniprot.org/)
- In the 'UniProtKB' query box, type Tfu_0901
- Click the 'Search' button
- The search returns one entry Q47RH8 (this is the UniProt accession number for Tfu_0901)
- Click the entry 'Q47RH8'
- Under the 'Sequence' section, you will find the protein sequence
- To retrieve the DNA sequence, go to 'Sequence database', click 'AAZ54939.1' (this is the GenBank accession number for Tfu_0901), and select 'FASTA'

UniProt also presents a summary of the protein sequence, which includes some very useful information for protein engineers. Navigate through the left sidebar:

- [PTM/Processing] There is a signal peptide (1–36) in the protein sequence of Tfu_0901 (we will discuss signal peptide in greater detail in ▶ Sect. 2.2).
- [Structure] The 3D structure of Tfu_0901 is available.
- [Cross-references] Under the section on 'Protein family/group databases', the protein belongs to glycoside hydrolase family 5 (GH5) in CAZy.

Another quick and easy way to find a specific protein sequence is to utilise the Protein Database of NCBI:

- Go to the Protein Database of NCBI (▶ https://www.ncbi.nlm.nih.gov/protein)
- In the 'Protein' query box, type Tfu_0901
- Click the 'Search' button
- You will find the protein sequence in the result page
- To view the protein sequence in FASTA format, click 'FASTA' on the third line of the header
- To retrieve the DNA sequence, click 'CDS' and select 'FASTA'

At this point, you may ask 'What is FASTA?'. FASTA format is a text-based format for representing either nucleotide sequences or protein sequences, in which bases or amino acids are represented using single-letter codes. A sequence in FASTA format begins with a single-line description, followed by lines of sequence data. The description line (defline) is distinguished from the sequence data by a greater-than ('>') symbol at the beginning. An example is given in ◻ Fig. 2.1. You can use a word processor (*e.g.*, Microsoft Word) and save the sequence as plain text (*.txt). A lot of bioinformatics tools for gene analysis (▶ Sect. 2.2) accept FASTA as the input type. Therefore, students are advised to understand the FASTA format before proceeding further. On top of that, we encourage students to use a monospaced font (also called a fixed-pitch, fixed-width, or non-proportional font) when displaying a protein sequence or a nucleotide sequence. Using these fonts (*e.g.*, Courier, Apercu Mono, GT Pressura Mono, Inconsolata, Maison Mono, Space Mono, Pitch, Roboto Mono, GT America Mono, and Akkurat Mono *etc*), all letters and characters will each occupy the same amount of horizontal space. Alignment of the letters makes it easier to compare different sequences visually. At this point, we would like to highlight a tool called EMBOSS seqret (▶ https://www.ebi.ac.uk/Tools/sfc/emboss_seqret/), which allows for conversion of protein, DNA and RNA sequences to various formats including FASTA.

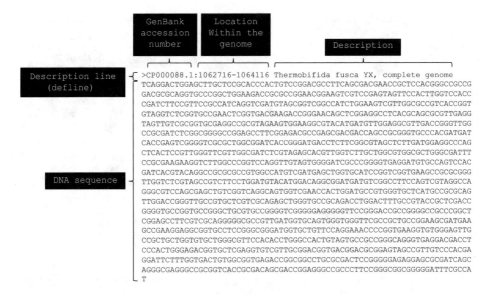

Fig. 2.1 The antisense strand of Tfu_0901 gene sequence is presented in a FASTA format. The coding strand would be the reverse complement of this gene sequence (see ▶ Sect. 2.2.2.1).

2.2 Gene Analysis

Now that you know how to retrieve a gene sequence, let's analyse the sequence. In this section, you will learn how to perform protein sequence analysis and DNA sequence analysis using our favourite tools in ☐ Table 2.2. Most of the tools in ☐ Table 2.2 can also be found in:

- The SIB Bioinformatics Resource Portal (ExPASy; ▶ https://www.expasy.org/)
- The European Bioinformatics Institute of EMBL (EMBL-EBI) webpage (▶ https://www.ebi.ac.uk/)

ExPASy provides free access to the scientific databases and software tools in different areas of life sciences, including proteomics, genomics, phylogeny, systems biology, population genetics, transcriptomics *etc*. To give you a brief guide on how to use the portal, let's search for a tool called BLAST:

- Go to the ExPASy webpage (▶ https://www.expasy.org/)
- In the left sidebar, click 'proteomics' under 'Categories'
- Click 'BLAST' under 'Tools' (all the tools are arranged in alphabetical order)

Similar to Swiss Institute of Bioinformatics (SIB), EMBL-EBI develops databases, tools, and software to align, verify and visualise the diverse data produced in publicly-funded research, and makes that information freely available to all. Using BLAST as an example again, let's explore the EMBL-EBI webpage:

- Go to the EMBL-EBI webpage (▶ https://www.ebi.ac.uk/)
- In the navigation menu, click 'Services'
- Click 'BLAST [protein]' under 'Tools'

2

◻ **Table 2.2**   Bioinformatics tools for protein sequence and DNA sequence analyses

Type of analysis	Tool	URL	When to use this tool?
Protein sequence analysis	ProtParam	▶ https://web.expasy.org/protparam/	Computes various physical and chemical parameters for a protein sequence
	SignalP	▶ http://www.cbs.dtu.dk/services/SignalP/	Predicts the presence of a signal peptide and the location of its cleavage site in a protein sequence
	BLAST	▶ https://blast.ncbi.nlm.nih.gov/Blast.cgi	Finds regions of similarity between biological sequences (protein or nucleotide sequence)
	Clustal Omega	▶ https://www.ebi.ac.uk/Tools/msa/clustalo/	Aligns multiple sequences (protein or nucleotide sequence)
	Pfam	▶ https://pfam.xfam.org/	Identifies protein family, domain(s) and functional site(s)
	PROSITE	▶ https://prosite.expasy.org/	Identifies protein family, domain(s) and functional site(s)
	SMART	▶ http://smart.embl-heidelberg.de/	Analyses domain architecture
	InterPro	▶ http://www.ebi.ac.uk/interpro/search/sequence/	Identifies protein family, domain(s) and functional site(s)
DNA sequence analysis	Reverse Complement	▶ https://www.bioinformatics.org/sms/rev_comp.html	Converts a DNA sequence into its reverse, complement, or reverse-complement counterpart
	DNA Stats	▶ https://www.bioinformatics.org/sms2/dna_stats.html	Returns the number of occurrences of each base in a DNA sequence
	Translate	▶ https://web.expasy.org/translate/	Translates a nucleotide (DNA/RNA) sequence to a protein sequence
	Codon Usage	▶ https://www.bioinformatics.org/sms2/codon_usage.html	Returns the number and frequency of each codon type in a DNA sequence
	Reverse Translate	▶ https://www.bioinformatics.org/sms2/rev_trans.html	Backtranslates a protein sequence into a nucleotide sequence

```
>AAZ54939.1 Cellulase. Glycosyl Hydrolase family 5 [Thermobifida fusca YX]
MAKSPAARKGGPPVAVAVTAALALLIALLSPGVAQAAGLTATVTKESSWDNGYSASVTVRNDTSSTVSQW
EVVLTLPGGTTVAQVWNAQHTSSGNSHTFTGVSWNSTIPPGGTASFGFIASGSGEPTHCTINGAPCDEGS
EPGGPGGPGTPSPDPGTQPGTGTPVERYGKVQVCGTQLCDEHGNPVQLRGMSTHGIQWFDHCLTDSSLDA
LAYDWKADIIRLSMYIQEDGYETNPRGFTDRMHQLIDMATARGLYVIVDWHILTPGDPHYNLDRAKTFFA
EIAQRHASKTNVLYEIANEPNGVSWASIKSYAEEVIPVIRQRDPDSVIIVGTRGWSSLGVSEGSGPAEIA
ANPVNASNIMYAFHFYAASHRDNYLNALREASELFPVFVTEFGTETYTGDGANDFQMADRYIDLMAERKI
GWTKWNYSDDFRSGAVFQPGTCASGGPWSGSSLKASGQWVRSKLQS
```

◘ **Fig. 2.2**   The protein sequence of Tfu_0901, presented in a FASTA format.

Many tools in ◘ Table 2.2 are so widely used that you will be able to find them even on common search engines like Google. We will use the protein sequence (◘ Fig. 2.2) and the DNA sequence (◘ Fig. 2.1) of Tfu_0901 as examples to illustrate the use of these tools. Most bioinformatics tools have user-friendly web-based interface, as long as you pay attention to the following points:

- Most tools accept a sequence in FASTA format or an accession number as input.
- If the FASTA format is not accepted, simply remove the description line.
- If a tool accepts the FASTA format but your input creates an error, check that there is an enter ↵ at the end of your definition line.
- Some tools provide either a help function or an example sequence to assist users.
- A typical error is the use of unauthorised characters in the input sequence. Therefore, it is important to clean your sequence up. Freeware such as the Sequence Cleanup or Formatting (► http://reverse-complement.com/cleanup.html) can be used to accomplish that.
- Be patient! Your job might be in a queue, which takes time to complete.
- For some tools, you can provide your email address. You will get an email notification when your job is completed.
- Most tools would also provide an explanation to the output.

2.2.1 Protein Sequence Analysis

Let's begin with protein sequence analysis. We will go through each of the tools in ◘ Table 2.2 in the order listed, by showing you the steps involved and the key information you can get out of each analysis.

2.2.1.1 ProtParam
Key steps:

- Go to the ProtParam webpage (► https://web.expasy.org/protparam/)
- Paste the protein sequence of Tfu_0901 MAK ...... LQS in the query box or type Q47RH8 in the accession number/sequence identifier search field
- Click 'Compute parameters' button

Important information derived from this analysis:

- Tfu_0901 is 466 amino acids long. The molecular weight of this endoglucanase is 49807.27 Da. → Useful when analysing this protein on an SDS-PAGE.

2

- The pI value is 5.21. → Useful for determining the pH of the protein purification buffers. pH should ideally be close to neutral, but 1–2 pH units away from the pI value. In the case of Tfu_0901, a pH of 7.5–8.0 is recommended.
- The molar extinction coefficient of this protein is 93850 M^{-1} cm^{-1}, assuming all cysteine residues are reduced. → Useful for estimating the concentration of purified protein by applying the Lambert-Beer law equation (C = protein concentration in M, A_{280} = absorbance at 280 nm, ε = molar extinction coefficient in M^{-1} cm^{-1}, l = pathlength in cm):

$$C = \frac{A_{280}}{\varepsilon \times l}$$

2.2.1.2 SignalP

Key steps:
- Go to the SignalP webpage (▶ http://www.cbs.dtu.dk/services/SignalP/)
- Paste the protein sequence in FASTA format in the query box or upload a FASTA file (*.txt)
- Under the 'Organism group', tick 'Gram-positive' (*T. fusca* is a Gram-positive bacterium)
- Click the 'Submit' button

Important information derived from this analysis:
- A signal peptide (1–36) exists in this protein sequence. Tfu_0901 is transported by the Sec translocon. → When you clone this gene for recombinant protein expression, say in *Escherichia coli* or other expression systems, you would need to remove the first 36 amino acids.
- The signal peptide is cleaved between residue 36 and residue 37 (AQA—AG) by the signal peptidase I (Lep).

2.2.1.3 BLAST

Key steps:
- Go to the BLAST webpage (▶ https://blast.ncbi.nlm.nih.gov/Blast.cgi)
- Under 'Web BLAST', click 'Protein BLAST'
- Paste the protein sequence of Tfu_0901 MAK ...... LQS in the query box or type AAZ54939.1 that is the GenBank accession number
- Click the 'BLAST' button

Important information derived from this analysis:
- Tfu_0901 is similar to the cellulase from *Thermobifida cellulosilytica* (percentage identity = 80.64%), endo-β-1,4-glucanase from *Thermobifida alba* (78.90%), and cellulase from *Thermobifida halotolerans* (76.39%). If you want to see the sequence alignment between Tfu_0901 and cellulase from *T. cellulosilytica*, you can click on the link of 'cellulase family glycosylhydrolase [Thermobifida cellulosilytica]' (◘ Fig. 2.3). If you click the link of accession number 'WP_068752955.1', you will find the protein sequence of this cellulase from *T. cellulosilytica*.

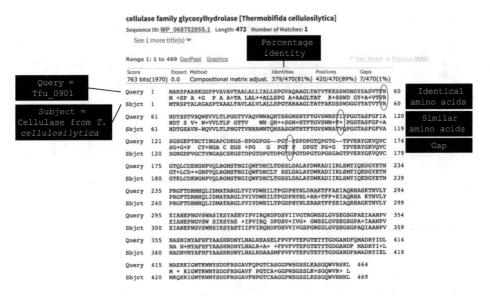

Fig. 2.3 Sequence alignment between Tfu_0901 and cellulase from *Thermobifida cellulosilytica*.

Important to point out, we have BLASTed Tfu_0901 against the database of 'Non-redundant protein sequences (nr)' in this demonstration. You can also BLAST against other databases to further refine your search, which is a useful feature. For instance, if you have isolated a new *Thermobifida* strain and have sequenced its genome, you can BLAST against this genome to find homologs of Tfu_0901 in this newly isolated organism.

Another useful feature is 'Align two or more sequences' in BLAST. If you want a quick alignment between two protein sequences to determine the percentage identity, this is your go-to tool. To give you a demonstration, let's align Tfu_0901 (accession number AAZ54939.1) and the cellulase from *Thermobifida cellulosilytica* (WP_068752955.1). Key steps involved are:

- Go to the BLAST webpage (▶ https://blast.ncbi.nlm.nih.gov/Blast.cgi)
- Under 'Web BLAST', click 'Protein BLAST'
- Tick 'Align two or more sequences'
- Paste the protein sequence of Tfu_0901 `MAK ...... LQS` in the first query box and the protein sequence of *T. cellulosilytica* cellulase `MTR ...... LAG` in the second query box or type `AAZ54939.1` in the first query box and `WP_068752955.1` in the second query box
- Click the 'BLAST' button

2.2.1.4 Clustal Omega

Unlike ProtParam, SignalP and BLAST that work on a single protein sequence input, Clustal Omega performs multiple-sequence alignment. For illustration purpose, let's align Tfu_0901 (accession number AAZ54939.1), cellulase from *T. cellulosilytica* (WP_068752955.1) and endo-β-1,4-glucanase from *Thermobifida alba* (BBA57842.1). Key steps:

- Go to the Clustal Omega webpage (▶ https://www.ebi.ac.uk/Tools/msa/clustalo/)
- Paste the sequences in FASTA format, as shown in ◼ Fig. 2.4, in the query box
- Click the 'Submit' button

2

```
>AAZ54939.1 Cellulase. Glycosyl Hydrolase family 5 [Thermobifida fusca YX]
MAKSPAARKGGPPVAVAVTAALALLIALLSPGVAQAAGLTATVTKESSWDNGYSASVTVRNDTSSTVSQW
EVVVLTLPGGTTVAQVWNAQHTSSGNSHTFTGVSWNSTIPPGGTASFGFIASGSGEPTHCTINGAPCDEGS
EPGGPGGPGTPSPDPGTQPGTGTPVERYGKVQVCGTQLCDEHGNPVQLRGMSTHGIQWFDHCLTDSSLDA
LAYDWKADIIRLSMYIQEDGYETNPRGFTDRMHQLIDMATARGLYVIVDWHILTPGDPHYNLDRAKTFFA
EIAQRHASKTNVLYEIANEPNGVSWASIKSYAEEVIPVIRQRDPDSVIIVGTRGWSSLGVSEGSGPAEIA
ANPVNASNIMYAFHFYAASHRDNYLNALREASELFPVFVTEFGTETYTGDGANDFQMADRYIDLMAERKI
GWTKWNYSDDFRSGAVFQPGTCASGGPWSGSSLKASGQWVRSKLQS

>WP_068752955.1 cellulase family glycosylhydrolase [Thermobifida cellulosilytica]
MTRSPTALRGASPTAAALTAVLALVLALLSPGTARAAGLTATFAKDSSWDGGYTATVTVRNDTGSAVNWQ
VVLTLPNGTTVNNAWNTQHSASGNTHTFTGVSWNATVQPGGTASFGFVASGNGDPVGCTVNGASCSEGST
DPGTDPGTDPGTDPGTDPGTDPGSGAGTPVERYGKVQVCGTKLCDKNGNPVQLRGMSTHGIQWFDHCLTG
SSLDALAYDWKADIIRLSMYIQEDGYETNPRGFTDRMHQLIDMATARGLYVIVDWHILTPGDPHYNLERA
RTFFSEIAQRHAGKTNVLYEIANEPNGVSWNSIKSYAETIIPVIRQHDPDSVVIVGSPGWSSLGVSEGSG
PAQIAANPVNADNVMYAFHFYAASHRDNYLNALRDAASMFPVFVTEFGTETYTGDGANDFAMADRYIELM
KQEKIGWTKWNYSDDFRSGAVFNPGTCAAGGPWSGSSLKSSGQWVRNHLLAG

>BBA57842.1 endo-1,4-beta glucanase E5 [Thermobifida alba]
MTRPPTALRGATPAAVALTAALALLLALLSPGVARAAGLTATFTKESTWDGGYTASVTVHNDTGGTVSHW
QVVVLTLPNGTTVNQAWNAQHTADGTSHTFTGVSWNSSIAPGGTASFGFTAGGSGDPVGCTVNGAPCAEGS
DPGTDPGGPGTDPGGPGTDPGGPGTDPGAGTPVERYGKVEVCGTKLCDENSNPVQLRGMSTHGIQWFDHC
LTDSSLDALAHDWQADIIRLSMYIQEDGYETNPRGFTDRMHQLIDMATARGLYVIVDWHILTPGDPHYNL
DRAKTFFAEIAQRHAGKENVLYEIANEPNGVSWASIKSYAEEVIPVTRQRDPDSVVIVGTPGWSSLGVSE
GSGPAEIAADPVDAGNVMYAFHFYAASHRDDYLDALRSAAQMFPVFVTEFGTESYTGDGTNDYVMTERYL
DLMKQEKIGWTKWNYSDDFRSGAVFTPGTCAAGGPWNGSSLKSSGQWARDQLRS
```

◻ **Fig. 2.4** Protein sequences (AAZ54939.1, WP_068752955.1, and BBA57842.1) used in a multiple-sequence alignment by Clustal Omega.

Important information derived from this analysis:
- As shown in ◻ Fig. 2.5, these 3 protein sequences have similar length (466–474 amino acids).
- Residues 121–159 (numbered according to Tfu_0901 or AAZ54939.1) is the region that shows highest variability.

Clustal Omega allows you to save the alignment file by clicking the 'Download Alignment File' button in the toolbar above the aligned sequences. You can also create a phylogenetic tree by clicking the 'Phylogenetic Tree' button.

2.2.1.5 Pfam

Key steps:
- Go to the Pfam webpage (▶ https://pfam.xfam.org/)
- Click 'SEQUENCE SEARCH' in the left sidebar
- Paste the protein sequence of Tfu_0901 MAK ...... LQS in the query box
- Click the 'Go' button

Important information derived from this analysis:
- Pfam identifies two domains within the protein sequence of Tfu_0901: a cellulase (glycosyl hydrolase family 5) domain (residues 179–431) and a cellulose binding domain (residues 39–139). → In some protein engineering studies, we want to investigate individual domain or create protein variants by fusing different domains together. Domain analysis is useful for this purpose.

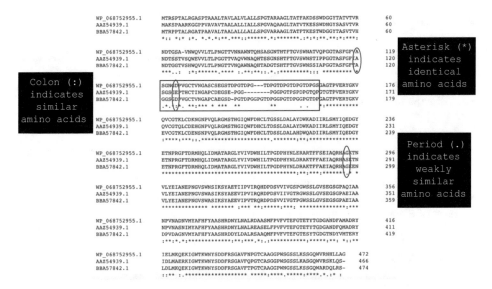

◻ Fig. 2.5 The output from a multiple-sequence alignment of AAZ54939.1, WP_068752955.1, and BBA57842.1 using Clustal Omega. Variable region is boxed in blue colour.

2.2.1.6 PROSITE

Key steps:
- Go to the PROSITE webpage (▶ https://prosite.expasy.org/)
- Paste the protein sequence of Tfu_0901 MAK …… LQS or type the UniProt accession number Q47RH8 in the query box
- Click the 'Scan' button

Important information derived from this analysis:
- PROSITE identifies a carbohydrate-binding type-2 domain (residues 32–139) within the protein sequence of Tfu_0901, which is consistent with the Pfam analysis above.

2.2.1.7 SMART

Key steps:
- Go to the SMART webpage (▶ http://smart.embl-heidelberg.de/)
- Type the UniProt accession number Q47RH8 or paste the protein sequence of Tfu_0901 MAK …… LQS in the query box
- Click the 'Sequence SMART' button

Important information derived from this analysis:
- SMART identifies a transmembrane region (residues 13–35) and a carbohydrate-binding domain (residues 39–136) within the protein sequence of Tfu_0901. This is in line with our previous SignalP (▶ Sect. 2.2.1.2), Pfam (▶ Sect. 2.2.1.5) and PROSITE (▶ Sect. 2.2.1.6) analyses.

2

2.2.1.8 InterPro

Key steps:
- Go to the InterPro webpage (▶ http://www.ebi.ac.uk/interpro/search/sequence/)
- Paste the protein sequence of Tfu_0901 MAK ...... LQS in the query box
- Click the 'Search' button
- Click the result link

Important information derived from this analysis:
- Consistent with the Pfam (▶ Sect. 2.2.1.5) result, InterPro identifies two domains within the protein sequence of Tfu_0901: a carbohydrate-binding type-2 domain (residues 32–139) and a glycoside hydrolase family 5 domain (179–429).
- In line with the SignalP (▶ Sect. 2.2.1.2) prediction, InterPro highlights the presence of a signal peptide (residues 1–36).

2.2.2 DNA Sequence Analysis

By now, we have covered many protein sequence analysis tools. Let's turn our attention to DNA sequence analysis. In ◻ Table 2.2, you might have noticed that we have cherry picked a few DNA analysis tools from the Sequence Manipulation Suite (SMS; ▶ https://www.bioinformatics.org/sms/). SMS is a collection of web-based programs for analysing and formatting DNA and protein sequences. Although many of the tools from SMS are not covered in this book, they are excellent tools. Therefore, we would strongly encourage you to browse through these tools in your own time.

2.2.2.1 Reverse Complement

◻ Figure 2.1 shows the antisense strand of Tfu_0901 gene sequence retrieved from the NCBI Protein Database. It is often useful to have the nucleotide sequence of the coding strand, a reverse complement of the antisense strand in the case of Tfu_0901. The Reverse Complement tool can convert the input DNA sequence to its reverse complement, reverse or complement counterpart.
  Key steps:
- Go to the Reverse Complement webpage (▶ https://www.bioinformatics.org/sms/rev_comp.html)
- Paste the DNA sequence of Tfu_0901 TCA ...... CAT into the query box
- Select 'reverse complement' in the dropdown menu
- Click the 'Submit' button

Important information derived from this analysis:
- The coding strand of Tfu_0901 in (5'→3') format

2.2.2.2 DNA Stats

Key steps:
- Go to the DNA Stats webpage (▶ https://www.bioinformatics.org/sms2/dna_stats.html)
- Paste the coding strand sequence of Tfu_0901 ATG ...... TGA into the query box
- Click the 'Submit' button

Important information derived from this analysis:

- The length of the gene is 1401 bp. This should be a number divisible by 3, as there are 466 amino acids and 1 stop codon. → Useful for DNA gel analysis after amplifying this gene in a PCR or after performing a restrictive digestion.
- The base composition is: A (18%), T (14%), G (29%) and C (39%). → This is a GC-rich gene. If we amplify this gene in a PCR, we might need a GC-rich buffer or a GC enhancer. We will discuss gene amplification in ▶ Sect. 5.1.3.

2.2.2.3 Translate

Although we already have the protein sequence of Tfu_0901, we want to demonstrate the Translate tool, as it may be useful for your other projects.
Key steps:

- Go to the Translate webpage (▶ https://web.expasy.org/translate/)
- Paste the coding strand sequence of Tfu_0901. ATG ...... TGA into the query box
- Click the 'Translate' button

Important information derived from this analysis:

- The nucleotide sequence is translated into a protein sequence.

2.2.2.4 Codon Usage

Key steps:

- Go to the Codon Usage webpage (▶ https://www.bioinformatics.org/sms2/codon_usage.html)
- Paste the coding strand sequence of Tfu_0901. ATG ...... TGA into the query box
- Click the 'Submit' button

Important information derived from this analysis:

- The frequency of each codon. For example, among the four codons for alanine (GCG, GCA, GCT, and GCC), three codons are used in the gene [*i.e.*, GCC (63%), GCG (33%) and GCT (4%)]. → Codon usage is an important determinant of protein expression yield. For instance, optimization of codon usage can potentially improve the expression of recombinant Tfu_0901 in *E. coli*. To do so, the next tool (Reverse Translate) might come in handy.

2.2.2.5 Reverse Translate

Key steps:

- Go to the Reverse Translate webpage (▶ https://www.bioinformatics.org/sms2/rev_trans.html)
- Paste the protein sequence of Tfu_0901. MAK ...... LQS into the query box
- Click the 'Submit' button

Important information derived from this analysis:

- A gene that is codon-optimized for heterologous protein expression in *E. coli*.

Although the default codon usage table is *E. coli*, this tool allows you to input any codon usage table of interest. The quickest way of finding the codon usage table of an organism is to visit the Codon Usage Database. Assuming that you want to express Tfu_0901 in *Bacillus subtilis* and you need the codon usage table of *B. subtilis*:

- Go to the Codon Usage Database webpage (▶ http://www.kazusa.or.jp/codon/)
- Type `Bacillus subtilis` in the query box
- Click the 'Submit' button
- Click the link of '*Bacillus subtilis* [gbbct]: 2529'

Most gene synthesis companies (*e.g.*, Eurofins, Integrated DNA Technologies, GenScript, GeneWiz, *etc*) will optimize the codon usage when designing a synthetic gene. As such, backtranslation tools are used less frequently these days.

2.2.3 **Perl**

Putting it simply, a protein or a DNA sequence is merely a string of characters. If you are familiar with Perl programming (▶ https://www.perl.org/), you may want to write your own scripts for simple sequence analyses or analyses you routinely perform. Perl is best known for its text processing capability – dealing with files, strings, and regular expressions. Important to point out, the BioPerl project (▶ https://bioperl.org/) contains many tools for bioinformatics, genomics, and life science written by Perl users and developers.

┌─ **Take-Home Messages** ───

1. Cellulose is a linear polysaccharide chain of repeating D-glucose units connected by $\beta(1{\rightarrow}4)$ glycosidic bonds.
2. Three types of enzyme are required for a complete cellulose hydrolysis to glucose, which include endo-β-1,4-glucanases, exoglucanases, and β-1,4-glucosidases.
3. The Kyoto Encyclopedia of Genes and Genomes (KEGG) and the National Center for Biotechnology Information (NCBI) are two excellent resources of gene information.
4. SnapGene Viewer is a useful tool for protein engineers. It allows visualization, annotation and sharing sequences.
5. The SIB Bioinformatics Resource Portal (ExPASy) and the European Bioinformatics Institute of EMBL (EMBL-EBI) webpage contain a host of bioinformatics tools for DNA sequence and protein sequence analysis.
6. FASTA format is a text-based format for representing either nucleotide sequences or protein sequences using single-letter codes. A sequence in FASTA format begins with a single-line description (defline), followed by lines of sequence data.
7. Perl is an excellent programming language for developing bioinformatics tools.

Exercise

(a) How many genes are annotated as β-1,4-glucosidases in the genome of *Thermobifida fusca* YX?

(b) What enzyme does Tfu_0937 encode?

(c) What reaction does Tfu_0937 catalyse?

(d) What is the pI value of Tfu_0937?

(e) Is there a signal peptide in the Tfu_0937 protein sequence?

(f) Is Tfu_0937 secreted by *Thermobifida fusca* YX?

(g) What is the GC content of the Tfu_0937 gene?

(h) Is the protein structure of Tfu_0937 available?

Further Reading

Artimo P, Jonnalagedda M, Arnold K, Baratin D, Csardi G, de Castro E, Duvaud S, Flegel V, Fortier A, Gasteiger E, Grosdidier A, Hernandez C, Ioannidis V, Kuznetsov D, Liechti R, Moretti S, Mostaguir K, Redaschi N, Rossier G, Xenarios I, Stockinger H (2012) ExPASy: SIB bioinformatics resource portal. Nucleic Acids Res 40(Web Server issue):W597–W603

Coordinators NR (2018) Database resources of the National Center for Biotechnology Information. Nucleic Acids Res 46(D1):D8–D13

Kanehisa M, Furumichi M, Tanabe M, Sato Y, Morishima K (2017) KEGG: new perspectives on genomes, pathways, diseases and drugs. Nucleic Acids Res 45(D1):D353–D361

Lombard V, Golaconda Ramulu H, Drula E, Coutinho PM, Henrissat B (2014) The carbohydrate-active enzymes database (CAZy) in 2013. Nucleic Acids Res 42(Database issue):D490–D495

Structural Analysis

Contents

© Springer Nature Switzerland AG 2020
T. S. Wong, K. L. Tee, *A Practical Guide to Protein Engineering*, Learning Materials in Biosciences,
https://doi.org/10.1007/978-3-030-56898-6_3

What You Will Learn in This Chapter

In this chapter, we will learn to:

- use Protein Data Bank, find and download relevant protein structures
- predict protein secondary structures
- generate protein models
- open *.pdb files
- use protein structure visualization tools

Proteins are intricate and fascinating molecular devices in the nanometer scale, where biological functions are exerted. Protein structural information provides insights into the structure-function relationship of a protein, and is an invaluable tool for protein engineers, especially for rational protein design. Before we continue, let's remind ourselves of the key properties that enable proteins to participate in a wide range of functions:

- Proteins are linear polymers or polypeptides, build from monomers called amino acids. The identity of a protein is determined by its amino acid sequence.
- The amino acid chain folds into a three-dimensional structure, held together by different bonds (*e.g.*, disulphide bond, electrostatic interaction, hydrophobic interaction, and van der Waals forces *etc*). Many proteins are modular, made from combinations of several identifiable, autonomously folding domains or modules.
- Proteins contain a wide variety of functional groups (*e.g.*, carboxylic acid, amine, thiol, and alcohol *etc*), and some proteins are post-translationally modified (*e.g.*, glycosylation, acetylation, biotinylation, and phosphorylation *etc*).
- Proteins can interact with one another and with other biological macromolecules (*e.g.*, DNA, RNA, and lipid *etc*) to form complex assemblies.
- Some proteins are rigid, whereas others display limited flexibility. Some proteins even contain intrinsically disordered regions.

Building on sequence analysis in ▶ Chap. 2, ▶ Chap. 3 will be dedicated to protein structure analysis.

3.1 Protein Structure Databases

From the protein sequence analysis of Tfu_0901 (▶ Sects. 2.1.1 and 2.1.4), we are made aware that its structural information is available (PDB codes of 2CKS and 2CKR). In this section, we will introduce how protein structural information can be retrieved from two important databases:

- Protein Data Bank (PDB; ▶ https://www.rcsb.org/)
- Biological Magnetic Resonance Data Bank (BMRB; ▶ http://www.bmrb.wisc.edu/)

3.1.1 Protein Data Bank (PDB)

Since 1971, PDB has served as the single repository of information about the 3D structures of proteins, nucleic acids, and complex assemblies. To search for and download the structure of Tfu_0901:

- Go to the PDB webpage (▶ https://www.rcsb.org/)
- Type 2CKS in the query box (you can also use 2CKR)
- Click the 'Search' button
- At the right upper corner of your results page, under the 'Download Files' drop-down menu, choose 'PDB Format'
- Save the 2cks.pdb file to your preferred location

Other than the structural information, PDB also provides a host of useful information for protein engineers. For example, as you scroll down the result page of 2CKS (▶ https://www.rcsb.org/structure/2CKS), the following information are provided:
- This endoglucanase can be recombinantly expressed using *B. subtilis*.
- If there is a journal publication associated with this structure, it will be listed under 'Literature'.
- Under 'Experimental Data & Validation', you will find details of crystallization, crystal diffraction, and diffraction data collection *etc.*

Protein Data Bank is easy to use when a PDB code is known, but not all proteins have structural information available. How would one know if structural information is available for your protein of interest? If the protein structure is available, how would you acquire its PDB code? To address these two questions, let's shift our attention to Tfu_1074, which is one of the endoglucanases in *T. fusca*. You should be able to retrieve the protein sequence of Tfu_1074 by now, using the various methods introduced in ▶ Sect. 2.1. In this section, we present four approaches to verify the existence of structural information and to retrieve the PDB codes for Tfu_1074:
1. Look up the gene Tfu_1074 in KEGG (▶ Sect. 2.1.1):
   - Go to the KEGG webpage (▶ https://www.genome.jp/kegg/)
   - In the 'KEGG' query box, type Tfu_1074
   - Click the 'Search' button
   - Under KEGG genes, click 'tfu:Tfu_1074'
   - Under entry 'Structure', you will find all the associated PDB codes (2BOG, 2BOE, 3RPT, 2BOD, 2BPF, and 1TML)
2. Look up the gene Tfu_1074 in UniProt (▶ Sect. 2.1.4)
   - Go to the UniProt webpage (▶ https://www.uniprot.org/)
   - In the 'UnitProtKB' query box, type Tfu_1074
   - Click the 'Search' button
   - Click entry 'Q47R05'
   - Under 'Structure', click 'Q47R05' and you will be directed to a page summarizing its structural information (PDB codes 2BOG, 1TML, 2BOE, 3RPT, 2BOD, and 2BOF)
3. Use the BLAST tool (▶ Sect. 2.2.1.3)
   - Go to the BLAST webpage (▶ https://blast.ncbi.nlm.nih.gov/Blast.cgi)
   - Under 'Web BLAST', click 'Protein BLAST'
   - Paste the protein sequence of Tfu_1074 MSP ...... AAS in the query box
   - In the 'Database' dropdown menu, choose 'Protein Data Bank proteins (pdb)'
   - Click the 'BLAST' button
   - The result entries showing high percentage identity will give you the associated PDB codes (1TML and 2BOE)
   - Check the description to confirm its relevance

3

4. Search in the PDB using keywords, identical to how you would do a literature search in PubMed (▶ Sect. 1.2):
   ▬ Go to the PDB webpage (▶ https://www.rcsb.org/)
   ▬ Type endoglucanase Thermobifida fusca in the query box
   ▬ Click the 'Search' button
   ▬ Identify relevant structures by checking the title and the protein sequence

You might wonder why some proteins have more than one PDB code. Each deposited structure in the repository will carry its own unique PDB code. The different codes represent independently obtained protein structures for the same protein (*e.g.,* deposited by different research groups). They can also be structures of the same protein crystallized with different ligands or they can be mutated forms of the same protein. For instance, both PDB codes 2CKS and 2CKR provide structural information for Tfu_0901, but 2CKR shows the structure of a Tfu_0901 variant carrying the mutation E391Q.

3.1.2 Biological Magnetic Resonance Data Bank (BMRB)

BMRB is a repository for data from NMR spectroscopy on proteins, peptides, nucleic acids, and other biomolecules. In term of number of deposits, this data bank is smaller in comparison to PDB. But, let's attempt to find some information from BMRB that would help Case Study 1:
▬ Go to the BMRB webpage (▶ http://www.bmrb.wisc.edu/)
▬ In the 'Title' query box, type Endoglucanase
▬ Click the 'Search' button
▬ You will get a list of three entries with BMRB IDs of 4148, 4706, and 27314 (as of 27th April 2020)

Although these search results are not specific to endoglucanases from *T. fusca*, they provide the solution structures of cellulose-binding domain and catalytic domain of endoglucanases, as well as how they interact with cello-oligosaccharides. Similar to PDB, BMRB provides a lot of useful information:
▬ PDB ID of the structure
▬ The associated journal publication (if there is)
▬ Host organism used for recombinant protein expression
▬ NMR data (*e.g.*, chemical shift)

3.2 Protein Structure Prediction

So far, we have only looked at proteins where the structural information is available. If a protein structure is unavailable (*e.g.*, Tfu_0937, one of the β-1,4-glucosidases in *T. fusca*), could we deduce structural information from the protein sequence? Well, the answer is yes. In this section, we will cover two topics:
▬ Protein secondary structure prediction
▬ Protein modelling

3.2.1 **Protein Secondary Structure Prediction**

The most common types of secondary structure in proteins are the α-helixes and the β-sheets. These secondary structures arise from the hydrogen bonds formed between the atoms of the polypeptide backbone. There are many bioinformatics tools that predict the secondary structures of proteins. In our experience, secondary structure prediction is particularly useful in situations when:

- We need to determine if a point mutation introduced would disrupt the secondary structure
- We need to define the protein domain boundaries

Let's now focus on three of the tools available (*i.e.*, JPred, PSIPRED, and PredictProtein), which have dedicated webservers and are easy to use. Since the structure of Tfu_0937 is unavailable, it is a good example for assessing the prediction capability of these tools. First, try to retrieve the protein sequence of Tfu_0937 using what you've learned in ► Sect. 2.1.

3.2.1.1 **JPred**

Key steps:

- Go to the JPred webpage (► http://www.compbio.dundee.ac.uk/jpred/)
- Paste the protein sequence of Tfu_0937 MTS ...... GQE in the query box
- Click the 'Make Prediction' button
- The programme will return a list of similar sequences with known protein structures
- Click the 'Continue' button

3.2.1.2 **PSIPRED**

Key steps:

- Go to the PSIPRED Workbench webpage (► http://bioinf.cs.ucl.ac.uk/psipred/)
- Paste the protein sequence of Tfu_0937 MTS ...... GQE in the query box
- Enter a job name Tfu_0937
- Click the 'Submit' button

3.2.1.3 **PredictProtein**

Key steps:

- Go to the PredictProtein webpage (► https://open.predictprotein.org/)
- Paste the protein sequence of Tfu_0937 MTS ...... GQE in the query box
- Click the 'PredictProtein' button

In ◘ Fig. 3.1, the predictions made by JPred, PSIPRED, and PredictProtein are summarized. As expected, the predictions are largely consistent across the three tools.

3.2.2 **Protein Tertiary Structure Prediction**

Generally, there are two approaches to predicting the tertiary structure of a target protein:

- Template-based modelling in which a previously determined structure of a related protein is used as a template to construct the atomic model of the target protein

3

```
AA               MTSQSTTPLGNLEETPKPDIRFPSDFVWGVATASFQIEGSTTADGRGPSIWDTFCATPGKVENGDTGDPACDHYNRYRDDVALMRELGVGAYRFSIAWPRIQPEGKGTPVEAGLDFYDRLVDCLLEAGIEFWPTLYHWDLPQALEDAG
JPred            -----EEEE----EE-----------------EEEEE-------------------HHHHHHHH-------EEE------HHHHHHHHHHHH------EEE-----HHHHH----
PSI PRED         ----------HHHHHH------------------HHHHHHHH-----EEE-------HHHHHHHHHHHHHHH------HHHHHH----
Predict Protein  ------EEEHHHHHHH----------HHHHH------HHHHHHHHHHHHHH----EEEEEEEEE------HHHHHHHHHHHHHHH----EEEEEE-----HHHHH----
SWISS-MODEL      -----EEEE-HHHH---------HHH-------HHHHHHHHHH-HHH----HHHHHHHHHH-HHH----EEEEEE------HHHHH----

AA               GWFNRDTAKRFADYAEIVYRRLGDRITNWNTLNEPWCSAFLGYASGVHAPGRQEPAAALAAAHHLMLGHGLAAAVMRDLAGQAGRSVRIGVAHNQTTVRPYTDSEADRDAARRIDALRNRIFTEPLVKGRYPEDLIEDVAAVTDYSFV
JPred            ---HHHHHHHHHHHHHHH-----EEE----EE-------------HHHHHHHHHHHHHHH--------EEEEE-----HHHHHHHHHH----
PSI PRED         ---HHHHHHHHHHHHHH-----EEE----HHHHH-HH-------HHHHHHHHHHHHHHH------EEEEE------HHH--HHHHHHHHH-
Predict Protein  --HHHHHHHHHHHHHHH-----EEEEE----EEEE------HHHHHHHHHHHHHHHHH-----EEEEEE-----HHHHHHHHHHHHHHH---EE---HHHHHHHHHH-
SWISS-MODEL      ---HHHHHHHHHHHHHH-------HHHHHHH------HHHHHHHHHHHHHHHHH------EEEE-----HHHHHHHHHHHHHH---HHHHHHHHH-

AA               QDGDLKTISANLDMMGVNFYNPSWVSGNRENGGSDRLPDEGYSPSVGSEHVEVDPGLPVTAMGWFIDPGLYDTLTRLANDYPGLPLYITENGAAFEDKVVDGAVHDTERIAYLDSHLRAAHAAIEAGVPLKGYFAWSFMDNFEWAL
JPred            -HHHHHHH-----HHHHH-----EEE---------------------------------E--HHHHHHHHHHH-----EEE-----HHHHH-
PSI PRED         -HHHHHH------EEEEE----HHEE---------HHHHHHHHHHHH-----EEE------HHHHHHHHHHHHHH------EEEE-----HHHHHHH-
Predict Protein  -HHHHHHH-----EEEEE-----EEEE-----HHHHHHHHHHHHHHH----EEEE------HHHHHHHHHHHHHHH---EEEE-HHHHHHHHHH-
SWISS-MODEL      -HHHHHH----EE--HHHHHHHHHH-----EEEE------HHHHHHHHHHHHHHH----EEEEEEEE--EE----HHHHHHHHHHHHHHHH---EEEE-----HHH-

AA               GYGKRFGIVHVDYESQTRTVKDSGWWYSRVMRNGGIFGQE
JPred            ------EEEEE------HHHHHHHHHHHHH----
PSI PRED         ----EEEE------HHHHHHHHHHHH----
Predict Protein  ---EEEEEEE------HHHHHHHHHHH----
SWISS-MODEL      ------EEE-----HHHHHHHHHHHHHH----
```

Fig. 3.1 Secondary structure prediction of Tfu_0937 using three bioinformatics tools (JPred, PSIPRED and PredictProtein). Amino acid (AA) sequence of Tfu_0937 is shown in blue. Symbols 'E', 'H', and '-' represent strand, helix, and other secondary structures, respectively. The predictions are compared to the secondary structure information derived from the homology model of Tfu_0937, obtained via SWISS-MODEL.

- Template-free modelling which relies on large-scale conformational sampling and the application of physics-based energy functions

3.2.2.1 Template-Based Modelling

The key steps in a standard template-based modelling include (a) selection of a suitable structural template, (2) alignment of the target sequence to the template structure, and (3) molecular modelling. Of all the tools developed for template-based modelling, we highly recommend SWISS-MODEL. It is a fully automated protein structure homolog modelling server that is very easy to use. To demonstrate how SWISS-MODEL works, let's try to create a model for Tfu_0937:

- Go to the SWISS-MODEL webpage (▶ https://swissmodel.expasy.org/)
- Click 'Start Modelling' button
- Paste the protein sequence of Tfu_0937 MTS ...... GQE or type Q47RE2 (the UniProt ID of Tfu_0937) in the query box
- Click the 'Build Model' button

To construct the model of Tfu_0937 (◘ Fig. 3.2a), SWISS-MODEL has selected the structure of β-glucosidase from *Streptomyces* sp. (PDB code 1GON) as a template. This is guided by the high sequence identity between Tfu_0937 and 1GON (◘ Fig. 3.2b). Not surprising, the secondary structure information derived from this homology model is in agreement with the predictions made by JPred, PSIPRED, and PredictProtein (◘ Fig. 3.1).

3.2.2.2 Template-Free Modelling

There are 5 key steps in a template-free modelling: (a) construct a multiple-sequence alignment, (b) predict local structure, (c) predict residues contacts, (d) assemble 3D models, and (e) refine and rank models. DMPFold, which is easy to use, belongs to the template-free modelling toolbox. Let's create another model for Tfu_0937 using DMPFold:

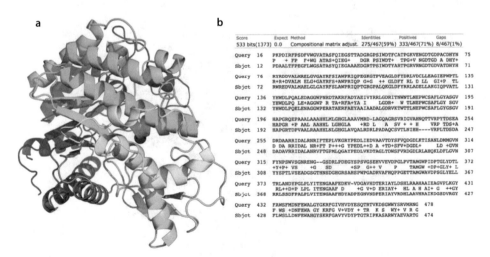

◘ Fig. 3.2 **a** Homology model of Tfu_0937, created using SWISS-MODEL with 1GON as a template. **b** Sequence alignment between Tfu_0937 (query) and 1GON (subject) using BLAST.

3

- Go to the PSIPRED Workbench webpage (▶ http://bioinf.cs.ucl.ac.uk/psipred/)
- Under 'Structure Modelling', tick 'DMPFold 1.0 Fast Mode'
- Paste the protein sequence of Tfu_0937 MTS …… GQE in the query box
- Enter a job name Tfu_0937
- Provide your email address
- Click the 'Submit' button
- You will get an email notification when the job is completed (the job would normally take hours to complete)

As is evident from ◧ Fig. 3.3, there are clear differences in the Tfu_0937 models created by SWISS-MODEL and DMPFold. Nonetheless, most of the helixes are consistent between the two models. Since Tfu_0937 is highly similar to 1GON (◧ Fig. 3.2b) in both protein sequence and protein function, the model created by SWISS-MODEL is likely to be more accurate.

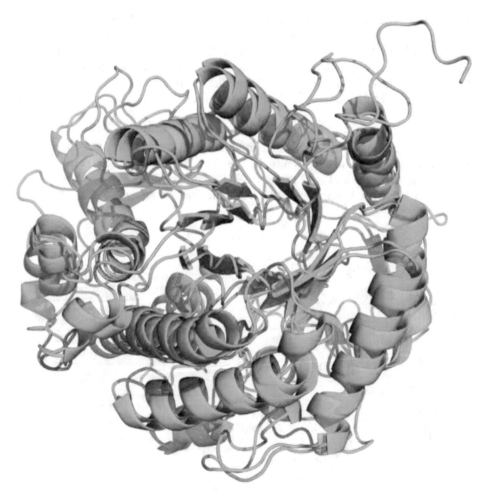

◧ **Fig. 3.3** Superimposition of the models of Tfu_0937, created using SWISS-MODEL (green) and DMPFold (cyan).

3.3 Protein Structure Visualization Tools

By this point, you should be confident in retrieving known protein structural information using their PBD code and performing structure prediction for proteins with no structural information. Some of the technical questions you might have at this point:
- How do I open a PDB file (*.pdb)?
- How do I view the models generated by SWISS-MODEL and DMPFold, which are saved as PDB files?
- How do I create the structural images in ◼ Figs. 3.2 and 3.3?
- How do I superimpose two protein structures, as shown in ◼ Fig. 3.3?

To address these questions, we would like to highlight a few protein structure visualization tools, which are great fun to learn, use and explore (◼ Table 3.1)! Each of these tools has its own dedicated webpage, populated with user guide, tutorials, and videos *etc*. Therefore, we would not repeat the information in this book. Instead, we would offer a couple of tips:
- If you struggle, Google it!
- There are loads of useful YouTube videos with step-by-step instructions!

◼ **Table 3.1** Protein structure visualization tools

Protein structure visualization tools	URL	Other useful online resources	System compatibility
PyMOL	▶ https://pymol.org/2/	PyMOL Wiki (▶ https://pymolwiki.org/index.php/Main_Page)	Windows, macOS, Linux
Swiss-PdbViewer (or DeepView)	▶ https://spdbv.vital-it.ch/		Windows, macOS
UCSF Chimera	▶ https://www.cgl.ucsf.edu/chimera/		Windows, macOS, Linux

---

Take-Home Messages

1. Protein function is determined by both its sequence and structure.
2. Protein structural information can be retrieved from Protein Data Bank (PDB) and Biological Magnetic Resonance Data Bank (BMRB).
3. PDB files have an extension of .pdb. The PDB format provides a standard representation for macromolecular structure data derived from X-ray diffraction and NMR studies.
4. The most common types of secondary structure in proteins are the α-helixes and the β-sheets. A range of bioinformatics tools is available for protein secondary structure prediction.

5. There are two approaches to protein tertiary structure prediction: template-based modelling and template-free modelling.
6. SWISS-MODEL is a fully automated protein structure homolog modelling server.
7. A range of protein structure visualization tools are made publicly available, and they are compatible with various operating systems.

Exercise

Case Study 3

The coronavirus (SARS-CoV-2) is a novel and highly pathogenic virus that emerged in 2019 from Wuhan, China. Its rapid international spread posed an unprecedented global public health emergency. The virus recognizes a receptor called angiotensin-converting enzyme 2 (ACE2) in human, through the receptor-binding domain (RBD) of its spike protein (S). A key to tackling the COVID-19 pandemic is to understand the receptor recognition mechanism of the virus, which regulates its infectivity, pathogenesis and host range. Your project supervisor would like you to express the full-length S and RBD recombinantly using an insect cell-based protein expression system.

(a) Find the protein sequence of SARS-CoV-2 RBD.
(b) Is the protein structure of SARS-CoV-2 RBD available?
(c) What protein expression system was used for the PDB entry 6M0J?
(d) How many disulphide bonds are there in SARS-CoV-2 RBD?
(e) Identify the cysteine pairs that form disulphide bonds in SARS-CoV-2 RBD.

Further Reading

Berman HM, Westbrook J, Feng Z, Gilliland G, Bhat TN, Weissig H, Shindyalov IN, Bourne PE (2000) The Protein Data Bank. Nucleic Acids Res 28(1):235–242

Waterhouse A, Bertoni M, Bienert S, Studer G, Tauriello G, Gumienny R, Heer FT, de Beer TAP, Rempfer C, Bordoli L, Lepore R, Schwede T (2018) SWISS-MODEL: homology modelling of protein structures and complexes. Nucleic Acids Res 46(W1):W296–W303

Protein Expression Hosts and Expression Plasmids

Contents

© Springer Nature Switzerland AG 2020
T. S. Wong, K. L. Tee, *A Practical Guide to Protein Engineering*, Learning Materials in Biosciences,
https://doi.org/10.1007/978-3-030-56898-6_4

4

What You Will Learn in This Chapter

In this chapter, we will learn to:
- choose an *E. coli* strain for cloning or protein expression purpose
- understand the genotype description of an *E. coli* strain
- search for a plasmid, retrieve its sequence, and place an order
- understand the information on a plasmid map
- create a plasmid map
- understand the key elements in a plasmid
- choose the right plasmid and the right *E. coli* host
- understand the use of a co-plasmid

In the previous three chapters, we have focused on dry lab research (*e.g.*, data search, analysis and modelling), which is instrumental to our subsequent experimental design and work. In this chapter, we shall transit into wet lab research, by looking at recombinant protein expression hosts and vectors. A variety of expression platforms have been used for recombinant protein production, including bacteria, yeasts, insect cells, mammalian cells, and fungi *etc* (◘ Table 4.1). It is practically impossible to cover all these expression systems in one book. As such, we shall focus our discussion on *E. coli*, which is arguably the most widely used expression system in protein engineering. Nonetheless, some of the principles and techniques covered in this book are directly applicable to other systems.

4.1 *Escherichia coli* Strains

E. coli is by far the most popular expression platform. Its use as a cell factory for protein production is well-established. The advantages of using *E. coli* include:
- Fast growth kinetics
- High cell density
- Inexpensive culture media
- Available genome information and models
- Large molecular biology toolbox available

◘ **Table 4.1**   Common expression hosts for recombinant protein production

Cell type	Category	Examples
Prokaryote	Bacteria	*Escherichia coli, Vibrio natriegens, Bacillus subtilis, Bacillus megaterium, Pseudomonas fluorescens, Pseudomonas putida, Corynebacterium glutamicum, Cupriavidus necator*
Eukaryote	Yeasts	*Saccharomyces cerevisiae, Pichia pastoris, Kluyveromyces lactis*
	Insect cells	Sf9 (*Spodoptera frugiperda*), High Five (*Trichoplusia ni*)
	Mammalian cells	Chinese hamster ovary cells (CHO), human embryonic kidney 293 cells (HEK-293)
	Fungi	*Aspergillus niger, Trichoderma reesei*

- Wide commercial availability
- Quick and easy transformation with exogenous DNA
- Vast volume of literatures and scientific data accumulated

Given our vast knowledge about this organism, it is not surprising that there is a long list of *E. coli* strains to choose from. The Coli Genetic Stock Center (CGSC) contains a huge collection of non-pathogenic laboratory strains of *E. coli*, which are widely used in genetic and molecular studies. Other than being an excellent *E. coli* database, *E. coli* strains in CGSC are also available upon request. CGSC also provides custom *E. coli* strain construction services. To search for a strain in CGSC (*e.g.*, DH5α), simply follow the steps below:

- Go to the CGSC webpage (▶ https://cgsc.biology.yale.edu/index.php)
- Click 'Strain Query'
- In the 'Strain Designation' query box, type DH5a
- Click the 'Submit' button
- You will find DH5α, along with its CGSC# of 14231 and genotype description

In ◘ Table 4.2, we summarized *E. coli* strains that you will most likely encounter whilst reading protein engineering literatures or working in a protein engineering laboratory. Some of these strains (*e.g.*, DH5α, TOP10F', XL1-Blue) are used mainly in routine cloning or subcloning, while others are dedicated to protein expression [*e.g.*, BL21, BL21(DE3), C41(DE3)]. If you are unfamiliar with bacterial genotype description, we have provided a guide to *E. coli* genetic markers in ◘ Table 4.3. This guide is by no means exhaustive. We have included genetic markers that we believe you will likely encounter to help you in choosing the right strain to do the right job!

For beginners, we recommend acquiring one strain for cloning/subcloning (DH5α) and two strains for protein expression [BL21(DE3) and C41(DE3)]. For instance, a single strain (DH5α) can be used for cloning, plasmid maintenance, and plasmid propagation. 'Keep things simple' is often our advice to students. In our experience, this small subset of strains would suffice for most protein engineering work.

4.2 Plasmid and Plasmid Map

As with the number of *E. coli* strains, there is a huge collection of plasmids that are compatible with *E. coli*. To help you navigate through this vast resource, we divide ▶ Sect. 4.2 into four parts to answer four questions that students commonly ask:

- Where do I find and order a plasmid?
- How do I read a plasmid map?
- How do I create a plasmid map?
- How to select the right plasmid?

4.2.1 Finding a Plasmid

Addgene is a non-profit global plasmid repository that archives and distributes plasmids for scientists. They also provide free molecular biology resources. To search and order a plasmid (*e.g.*, pUC19, a common cloning vector) from Addgene:

4

◼ **Table 4.2** Commonly used *E. coli* strains (arranged in alphabetical order), their genotypes and applications

E. coli strains	Genotypes	Application	
		Cloning	Expression
ArcticExpress (DE3)	F⁻ *ompT hsdS*_B (r_B^-, m_B^-) *dcm*⁺ Tet^R *gal* λ(DE3) *endA* Hte [*cpn10 cpn60* Gent^R]		•
BL21	F⁻ *ompT hsdS*_B (r_B^-, m_B^-) *gal dcm*		•
BL21(DE3)	F⁻ *ompT hsdS*_B (r_B^-, m_B^-) *gal dcm* (DE3)		•
BL21(DE3)pLysE	F⁻ *ompT hsdS*_B (r_B^-, m_B^-) *gal dcm* (DE3) pLysE(Cam^R)		•
BL21(DE3)pLysS	F⁻ *ompT hsdS*_B (r_B^-, m_B^-) *gal dcm* (DE3) pLysS(Cam^R)		•
BL21-CodonPlus(DE3)-RIPL	F⁻ *ompT hsdS*_B (r_B^-, m_B^-) *dcm*⁺ Tet^R *gal* λ(DE3) *endA* Hte [*argU proL* Cam^R] [*argU ileY leuW* Strep/Spec^R]		•
ClearColi BL21(DE3)	F⁻ *ompT hsdS*_B (r_B^-, m_B^-) *gal dcm lon* λ(DE3 [*lacI lacUV5-T7 gene 1 ind1 sam7 nin5*]) *msbA148 ΔgutQ ΔkdsD ΔlpxLΔlpxMΔpagPΔlpxP ΔeptA*		•
DH5α	F⁻ *φ80lacZΔM15 Δ(lacZYA-argF)U169 recA1 endA1 hsdR17*(r_K^-, m_K^+) *phoA supE44* λ⁻ *thi-1 gyrA96 relA1*	•	
DH10B	F⁻ *mcrA Δ(mrr-hsdRMS-mcrBC) φ80lacZΔM15 ΔlacX74 recA1 endA1 araD139 Δ(ara-leu)7697 galU galK* λ⁻ *rpsL*(Str^R) *nupG*	•	
HMS174(DE3)	F⁻ *recA1 hsdR*(r_{K12}^-, m_{K12}^+) (DE3) (Rif^R)		•
JM109	*endA1 recA1 gyrA96 thi hsdR17* (r_K^-, m_K^+) *relA1 supE44 Δ(lac-proAB)* [F′ *traD36, proAB, laqI^qZΔM15*]	•	
M15	Nal^S Str^S Rif^S *thi lac ara*⁺ *gal*⁺ *mtl* F⁻ *recA*⁺ *uvr*⁺ *lon*⁺		•
NovaBlue	*endA1 hsdR17*(r_{K12}^-, m_{K12}^+) *supE44 thi-1 recA1 gyrA96 relA1 lac* [F′ *proA*⁺*B*⁺ *lacI^qZΔM15*::Tn*10*]	•	
Origami™ 2(DE3)	*Δ(ara-leu)7697 ΔlacX74 ΔphoA PvuII phoR araD139 ahpC galE galK rpsL* F′[*lac*⁺ *lacI^q pro*] (DE3) *gor522*::Tn*10 trxB* (Str^R, Tet^R)		•
OverExpress C41(DE3)	F⁻ *ompT hsdS*_B (r_B^-, m_B^-) *gal dcm* (DE3)		•
OverExpress C41(DE3)pLysS	F⁻ *ompT hsdS*_B (r_B^-, m_B^-) *gal dcm* (DE3) pLysS(Cam^R)		•
OverExpress C43(DE3)	F⁻ *ompT hsdS*_B (r_B^-, m_B^-) *gal dcm* (DE3)		•

◨ **Table 4.2**   (continued)

E. coli strains	Genotypes	Application	
		Cloning	**Expression**
OverExpress C43(DE3)pLysS	F⁻ *ompT hsdS_B (r_B⁻, m_B⁻) gal dcm* (DE3) pLysS(Cam^R)		•
Rosetta™ 2(DE3)	F⁻ *ompT hsdS_B (r_B⁻, m_B⁻) gal dcm* (DE3) pRARE2 (Cam^R)		•
SG13009	Nal^S Str^S Rif^S *thi lac ara⁺ gal⁺ mtl* F⁻ *recA⁺ uvr⁺ lon⁺*		•
SHuffle® T7	F′ *lac pro lacI^q* / Δ*(ara-leu)7697 araD139 fhuA2 lacZ::T7 gene1 Δ(phoA) PvuII phoR ahpC* galE (or U) galK λatt::pNEB3-r1-cDsbC* (Spec^R, *lacI^q*) Δ*trxB rpsL150*(Str^R) Δ*gor* Δ*(malF)3*		•
TOP10	F⁻ *mcrA* Δ*(mrr-hsdRMS-mcrBC)* φ80*lacZ*Δ*M15* Δ*lacX74 recA1 araD139* Δ*(ara-leu)7697 galU galK* λ⁻ *rpsL*(Str^R) *endA1 nupG*	•	•
TOP10F′	F′ [*lacI^q*, Tn*10*(Tet^R)] *mcrA* Δ*(mrr-hsdRMS-mcrBC)* φ80*lacZ*Δ*M15* Δ*lacX74 recA1 araD139* Δ*(ara-leu)7697 galU galK rpsL(*Str^R*) endA1 nupG*	•	
XL1-Blue	*recA1 endA1 gyrA96 thi-1 hsdR17 supE44 relA1 lac* [F′ *proAB lacI^qZ*Δ*M15* Tn*10* (Tet^R)]	•	
XL10-Gold	Tet^R Δ*(mcrA)183* Δ*(mcrCB-hsdSMR-mrr)173 endA1 supE44 thi-1 recA1 gyrA96 relA1 lac* Hte [F′ *proAB lacI^qZ*Δ*M15* Tn*10* (Tet^R) Amy Cam^R]	•	

◨ **Table 4.3**   Guide to *E. coli* genetic markers (arranged in alphabetical order)

Marker	Description	Functional consequence
ara	Disruption of arabinose metabolism pathway.	Inability to utilize arabinose as a carbon source.
Cam^R		Confers resistance to chloramphenicol.
dam	Mutation in DNA adenine methylase (5′-GATC-3′).	Preparing unmethylated DNA, important when performing restriction digest with certain restrictive enzymes.
dcm	Mutation in DNA cytosine methylase (5′-CCWGG-3′).	Preparing unmethylated DNA, important when performing restriction digest with certain restrictive enzymes.
(DE3)	λ lysogen that encodes T7 RNA polymerase.	Induces expression in T7-driven expression systems.

(continued)

4

◻ **Table 4.3** (continued)

Marker	Description	Functional consequence
endA, *endA1*	Mutation in endonuclease I (nonspecific cleavage of dsDNA).	Improves plasmid yield.
F	Host does (F′) or does not (F⁻) contain the fertility plasmid.	A low copy-number plasmid, encoding proteins required for bacterial conjugation. Genes listed on F′ are wild-type unless indicated otherwise.
fhuA	Mutation in the ferric hydroxamate uptake, iron uptake receptor.	Confers T1/T5 phage resistance.
Gal	Mutation in the galactose metabolism pathway.	Galactose cannot be used as a carbon source for bacterial growth.
hsdRMS	Mutations in the methylation and restriction system genes. The genetic marker may contain the allele number (*e.g., hsdR17*) and phenotype [*e.g.,* (r_K^-, m_K^+)], where r is for restriction, m for methylation, and the subscript K for the parental strain *E. coli* K12).	Allows cloning of DNA from non-*E. coli* sources, such as PCR products, without cleavage by endogenous restriction endonucleases.
Hte		High transformation efficiency.
lacI^q	Overproduces the *lacI* gene product, a repressor of the *lac* operon.	Inhibits transcription from the *lac* promoter, which can be overcome by IPTG addition. Allows controlled gene expression from promoters that carry the *lac* operator.
lon	Mutation in the *lon* (ATPase-dependent protease) gene.	Reduces degradation of expressed recombinant proteins.
mcrA, *mcrBC*	Inactivation of pathway that cleaves methylated cytosine DNA.	Allows for uptake of foreign (methylated) DNA.
mrr	Inactivation of pathway that cleaves methylated adenine or cytosine DNA.	Allows for uptake of foreign (methylated) DNA.
nupG	Mutation in a nucleoside transport gene.	Increases plasmid uptake.
ompT	Mutation in the *ompT* (outer membrane protease) gene.	Reduces the degradation of expressed recombinant proteins.
pLys	Plasmid that encodes T7 lysozyme.	Inhibits binding of T7 RNA polymerase to the T7 promoter. Reduces basal expression of cloned genes driven by the T7 expression system.
recA, *recA1,* *recA13*	Mutation in a DNA-dependent ATPase that is essential for recombination and general DNA repair.	Reduces plasmid recombination, increases plasmid stability.
relA	Mutation in a regulatory gene for coupling between transcription and translation.	Allows RNA synthesis in limiting amino acid concentrations or in the absence of protein synthesis (*i.e.,* relaxed phenotype).
Tet^R		Confers resistance to tetracycline.

- Go to the Addgene webpage (▶ https://www.addgene.org/)
- Type pUC19 in the query box
- Click the 'Search' button
- Under 'Catalog', click the 'Plasmids' box
- You will find pUC19 with a catalog number of #500005
- Click on the 'pUC19' link
- To order, click 'Add to Cart'
- To download the plasmid sequence, click 'View all sequences' then click 'GenBank' or 'SnapGene'

If you have browsed through the CGSC webpage (▶ Sect. 4.1), you probably are aware that it is possible to request an *E. coli* strain that harbours a plasmid from CGSC:

- Go to the CGSC webpage (▶ https://cgsc.biology.yale.edu/index.php)
- Click 'Gene/Plasmid/Phage/Chrom.Sites Query'
- In the 'Name' query box, type pUC19
- Select 'Plasmid' in the 'Type' dropdown menu
- Click the 'Submit' button
- You will find pUC19, which has been transformed into DH5α (CGSC# 12555)

Commercial companies have dedicated significant effort into developing plasmids for cloning and expression. Examples of these plasmid systems include Novagen's pET vectors, Invitrogen's pBAD vectors, New England Biolabs' pMAL vectors and Promega's pGEM-T vectors. Commercial plasmids can be directly purchased from their suppliers.

4.2.2 Reading a Plasmid Map

A plasmid map is a graphical representation of a plasmid. The plasmid map of pUC19 is shown in ◘ Fig. 4.1 and we have captured the important information in ◘ Table 4.4. Reading a plasmid map is important for various reasons:

- Deciding on which plasmid to use.
- Handling of your plasmid when it arrives (*e.g.*, What antibiotic to use when you transform the plasmid into an *E. coli* strain?).
- Designing your cloning strategy.
- Selecting the right sequencing primer for clone verification.
- Using the right recombinant protein expression system.

4.2.3 Creating a Plasmid Map

The plasmid map of pUC19 in ◘ Fig. 4.1 is created using SnapGene, a software that we have introduced in ▶ Sect. 2.1.3. All the plasmid maps in Addgene are created using SnapGene. We therefore strongly encourage students to familiarize yourself with this software.

4

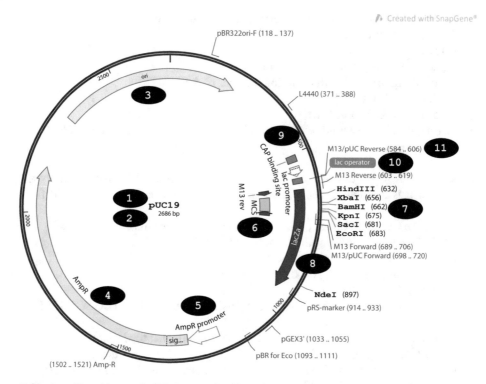

○ **Fig. 4.1**   Plasmid map of pUC19, created using SnapGene.

SnapGene opens both the GenBank files (*.gbk) and the SnapGene files (*.dna). GenBank file (○ Fig. 4.2) consists of an annotation section and a sequence section. The important data elements of a GenBank file are explained in ○ Table 4.5.

Comparing the plasmid map (○ Fig. 4.1) and the GenBank file (○ Fig. 4.2) of pUC19, you will notice that SnapGene is capable of 'reading' all the features in the GenBank file and presenting them in a graphical format.

4.2.4 Plasmid Selection

Generally, plasmids are divided into two major categories: cloning vector and expression vector. It is not practical to compile all the plasmids available. Instead, we have provided a list of plasmids that we often see in protein engineering literatures (○ Table 4.6). As you go through this table, it is useful to note that:

- Most of these plasmids are members of a much larger plasmid family. For instance, pET24a (+) is a member of the pET protein expression system. There are usually only minor differences between each member. Variations are typically

◼ **Table 4.4** Important information gathered from the plasmid map of pUC19

Number	Element	Description
1	pUC19	The plasmid name that always begins with a small 'p'.
2	2686 bp	The plasmid size.
3	ori	Origin of plasmid replication, which determines the copy number of the plasmid and its plasmid incompatibility group.
4	AmpR	β-Lactamase gene, which confers resistance to ampicillin.
5	AmpR promoter	The promoter of the β-lactamase gene.
6	MCS	A multiple cloning site (or a polylinker region) is a DNA region containing multiple unique restriction enzyme cut sites for cloning a target gene.
7	HindIII, XbaI, BamHI, KpnI, SacI, EcoRI, and NdeI	Unique restriction enzyme cut sites, written in boldface.
8	lacZα	The lacZ α-complementation gene (*lacZα*), which encodes the α-peptide, an inactive segment of the β-galactosidase. The presence of the *lacZα* gene in complementing *E. coli* strains (which contain a chromosomal copy of the remaining β-galactosidase gene fragment) results in the reconstitution of a functional β-galactosidase, giving blue colonies on media supplemented with X-gal.
9	*lac* promoter	The promoter of the *lacZα* gene (before gene cloning) and of the target gene (after gene cloning).
10	*lac* operator	An inhibitor protein (LacI) binds to this regulatory site (*lacO*) in the promoter and turns off transcription from the *lac* promoter.
11	M13/pUC Reverse and M13 Reverse	Potential sequencing primers to verify the clone via DNA sequencing.

found in the promoter, MCS, tag, tag position, tag cleavage site, selection marker, and the number of expression cassettes.

— Plasmids are sometimes engineered to suit certain applications (*e.g.*, add/remove a restriction site, add/remove a tag, add a linker, and alter the promoter/terminator *etc*). This creates many more variations of the plasmids in ◼ Table 4.6.

— When taking on an existing project, you often inherit a plasmid from someone else. In such situation, you may not need to select or create a new plasmid.

4

Annotation

①

```
LOCUS       Exported                2686 bp ds-DNA     circular SYN 13-SEP-2018
DEFINITION  pUC cloning vector.
ACCESSION   .
VERSION     .
KEYWORDS    .
SOURCE      synthetic DNA construct
  ORGANISM  synthetic DNA construct
REFERENCE   1  (bases 1 to 2686)
```

② ③

```
                        pUC19
  AUTHORS   Norrander J., Kempe T., Messing J.
  TITLE     Construction of improved M13 vectors using
            oligodeoxynucleotide-directed mutagenesis.
  JOURNAL   Gene. 1983 Dec;26(1):1101-6.
  PUBMED    6323249
REFERENCE   2  (bases 1 to 2686)
  AUTHORS   .
  TITLE     Direct Submission
  JOURNAL   Exported Sep 13, 2018 from SnapGene Server 1.1.58
            http://www.snapgene.com
FEATURES             Location/Qualifiers
```

④

```
     source          1..2686
                     /organism="synthetic DNA construct"
                     /mol_type="other DNA"
     primer_bind     158..175
                     /label=pBR322ori-F
                     /note="pBR322 origin, forward primer"
     primer_bind     371..388
                     /label=L4440
                     /note="L4440 vector, forward primer"
     protein_bind    505..526
                     /label=CAP binding site
                     /bound_moiety="E. coli catabolite activator protein"
                     /note="CAP binding activates transcription in the presence
                     of cAMP."
     promoter        541..571
                     /label=lac promoter
                     /note="promoter for the E. coli lac operon"
     protein_bind    579..595
                     /label=lac operator
                     /bound_moiety="lac repressor encoded by lacI"
                     /note="The lac repressor binds to the lac operator to
                     inhibit transcription in E. coli. This inhibition can be
                     relieved by adding lactose or
                     isopropyl-beta-D-thiogalactopyranoside (IPTG)."
     primer_bind     584..606
                     /label=M13/pUC Reverse
                     /note="In lacZ gene"
     primer_bind     603..619
                     /label=M13 rev
                     /note="common sequencing primer, one of multiple similar
                     variants"
     primer_bind     603..619
                     /label=M13 Reverse
                     /note="In lacZ gene. Also called M13-rev"
     CDS             632..688
                     /codon_start=1
                     /gene="lacZ fragment"
                     /product="LacZ-alpha fragment of beta-galactosidase"
                     /label=lacZ-alpha
                     /translation="MTMITPSLHACRSTLEDPRVPSSNSLAVVLQRBGWENPGVTQLMR
                     LAAHPPPASWRNSEEARTDPSQQLRSLNGEWRLMRYFLLTHLCGISHRIWCTLSTICS
                     DAA"
     misc_feature    632..688
                     /label=MCS
                     /note="pUC18/19 multiple cloning site"
     primer_bind     complement(689..706)
                     /label=M13 Forward
                     /note="In lacZ gene. Also called M13-F20 or M13 (-21)
                     Forward"
     primer_bind     complement(689..705)
                     /label=M13 fwd
                     /note="common sequencing primer, one of multiple similar
                     variants"
     primer_bind     complement(698..720)
                     /label=M13/pUC Forward
                     /note="In lacZ gene"
     primer_bind     complement(914..933)
                     /label=pRS-marker
                     /note="pRS vectors, use to sequence yeast selectable
                     marker"
```

Annotation

```
     primer_bind     1033..1055
                     /label=pGEX 3'
                     /note="pGEX vectors, reverse primer"
     primer_bind     complement(1093..1111)
                     /label=pBRforEco
                     /note="pBR322 vectors, upstream of EcoRI site, forward
                     primer"
     promoter        1179..1283
                     /gene="bla"
                     /label=AmpR promoter
     CDS             1284..2144
                     /codon_start=1
                     /gene="bla"
                     /product="beta-lactamase"
                     /label=AmpR
                     /note="confers resistance to ampicillin, carbenicillin, and
                     related antibiotics"
                     /translation="MSIQHFRVALIPFFAAFCLPVFAHPETLVKVKDAEDQLGARVGYI
                     ELDLNSGKILESFRPEERFPMMSTFKVLLCGAVLSRIDAGQEQLGRRIHYSQNDLVEYS
                     PVTEKHLTDGMTVRELCSAAITMSDNTAANLLLTIIGGPKELTAFLHNMGDHVTRLDRW
                     EPELNEAIPNDERDTTMPVAMATTLRKLLTGELLTLASRQQLIDWMEADKVAGPLLRSA
                     LPAGWFIADKSGAGERGSRGIIAALGPDGKPSRIVVIYTTGSQATMDERNRQIAEIGAS
                     LIKHW"
     primer_bind     complement(1502..1521)
                     /label=Amp-R
                     /note="Ampicillin resistance gene, reverse primer"
     rep_origin      complement(2317)
                     /direction=RIGHT
                     /label=ori
                     /note="high-copy-number ColE1/pMB1/pBR322/pUC origin of
                     replication"
```

Sequence

⑤

```
ORIGIN
        1 gagcaccga caagctgagc tatgagataca cgcacacgcct ccgaacgga gaaagcggaa
       61 caggtatccg gggtcggaac aggagacggc acgagggtgg ttccagggga
      121 aaaggcctgg tatcttatca gtcctgccac cctgacttg agggtcgatt
      ...
     2641 acgccagc ttggagggaa cgaccacaac cgaact
```

● **Fig. 4.2** GenBank file of pUC19.

◘ Table 4.5 Data elements in a GenBank file

Number	Data element	Explanation
1	Definition	A brief description of the sequence
2	Sequence length	Number of nucleotide base pairs
3	Molecule type	The type of molecule that was sequenced
4	Features	Information about genes and gene products, as well as regions of biological significance reported in the sequence.
5	Sequence	Nucleotide sequence

— Plasmid selection is often influenced by prior user experience and an existing working protocol.

For beginners, if you have to start from scratch, we would advise beginning with one of the plasmids listed in ◘ Table 4.6. These plasmids will suit most protein engineering projects. It is not difficult to spot the similarities among these plasmids:
— High copy number plasmid
— Typical plasmid size 3–6 kb
— Ampicillin and kanamycin are the two most common selection markers
— Most frequently used promoter is T7*lac*
— Typical tag for protein purification is 6×His
— Thrombin is frequently used for tag removal

We suggest applying the decision making survey outlined in ◘ Fig. 4.3 to help you identify your needs, and hence, the right plasmid to use.

You may have realised by now that the large number of plasmids available is attributed to variability in the three elements listed here:
— Promoter type, strength, and configuration
— Protein tag to enhance expression level, increase solubility, enable secretion, facilitate purification, and allow for labelling and detection *etc*
— Protease used to cleave the protein tag off

In the sections below, we will go through each of these elements in greater details.

4

◻ **Table 4.6** Commonly used plasmids for gene cloning and protein expression in *E. coli*

Plasmid	Cloning (C) / Expression (E)	Copy number	Length [bp]	Antibiotic selection marker	Promoter	Tag	Tag cleavage[a]	Single (S) / double (D) gene expression
pASG-IBA145	E	High	3973	Amp	Tet	*N*-Twin Strep Tag, *C*-6×His	Nil	S
pBAD/His A	E	High	4102	Amp	araBAD	*N*-6×His-Xpress Epitope	EK	S
pBluescript II SK (+)	C	High	2961	Amp	*lac*	N/A	N/A	N/A
pCDFDuet-1	E	High	3781	Sm	T7*lac*, T7*lac*	*N*-6×His, C-S Tag	Nil	D
pCOLADuet-1	E	High	3719	Kan	T7*lac*, T7*lac*	*N*-6×His, C-S Tag	Nil	D
pET24a (+)	E	High	5310	Kan	T7*lac*	*N*-T7 Tag, *C*-6×His	Nil	S
pET32a (+)	E	High	5900	Amp	T7*lac*	*N*-Trx Tag-6×His-S Tag	T	S
pET33b (+)	E	High	5383	Kan	T7*lac*	*N*-6×His-PKA Site-T7 Tag, *C*-6×His	T	S
pET35b (+)	E	High	5927	Kan	T7*lac*	*N*-CBD Tag-S Tag, *C*-6×His	T, X	S
pET39b (+)	E	High	6106	Kan	T7*lac*	*N*-DbsA-6×His - S Tag, *C*-6×His	T, EK	S
pETDuet-1	E	High	5420	Amp	T7*lac*, T7*lac*	*N*-6×His, C-S Tag	Nil	D
pET SUMO	E	High	5643	Kan	T7*lac*	*N*-6×His-SUMO	S	S
pGEM-T	C	High	3000	Amp	N/A	N/A	N/A	N/A
pGEX-3X	E	High	4952	Amp	taclac	*N*-GST	X	S
pMAL-c5X	E	High	5677	Amp	taclac	*N*-MBP	X	S
pPSG-IBA45	E	High	3532	Amp	T7	*N*-Strep Tag II, *C*-6×His	Nil	S
pPSG-IBA145	E	High	3592	Amp	T7	*N*-Twin Strep Tag, *C*-6×His	Nil	S

(continued)

◼ **Table 4.6** (continued)

Plasmid	Cloning (C) / Expression (E)	Copy number	Length [bp]	Antibiotic selection marker	Promoter	Tag	Tag cleavage[a]	Single (S) / double (D) gene expression
pQE-30	E	High	3461	Amp	T5*lac*	*N*-6×His	Nil	S
pRSET A	E	High	2897	Amp	T7	*N*-6× His-T7 Gene 10 Leader-Xpress Epitope	EK	S
pRSFDuet-1	E	High	3829	Kan	T7*lac*, T7*lac*	*N*-6× His, *C*-S Tag	Nil	D
pTrcHis2 A	E	High	4406	Amp	trc*lac*	*C-myc* Epitope-6×His	Nil	S
pUC18	C	High	2686	Amp	*lac*	N/A	N/A	N/A
pUC19	C	High	2686	Amp	*lac*	N/A	N/A	N/A
pUC57	C	High	2710	Amp	*lac*	N/A	N/A	N/A

[a]EK Enterokinase, T Thrombin, X Factor Xa

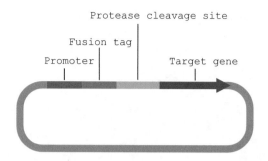

4

Colour	Question	Your options
	(1) What is your purpose?	☐ Cloning ☐ Protein expression
	(2) How many genes do you intend to clone?	☐ 1 gene ☐ 2 genes
	(3) What type of protein expression?	☐ Constitutive ☐ Inducible
	(4) What promoter do you intend to use?	☐ *ara*BAD ☐ *lac* ☐ *lac*UV5 ☐ *phoA* ☐ pL ☐ pR ☐ *rha*BAD ☐ Sp6 ☐ T3 ☐ T5 ☐ T7 ☐ T7 *lac* ☐ *tac* ☐ *tet* ☐ *trc* ☐ *trp*
	(5) Do you need to tag your protein?	☐ Yes ☐ No
	(6) What is the purpose of your tag?	☐ Protein purification ☐ Western blot / immunoprecipitation / immunochemistry ☐ Protein expression yield and solubility ☐ Labelling / detection ☐ Protein secretion
	(7) Do you have a preference on the tag position?	☐ N-terminus ☐ C-terminus
	(8) Do you need to cleave the tag off?	☐ Yes ☐ No
	(9) Do you have a preference on the protease used for tag cleavage?	☐ Enterokinase ☐ HRV-3C ☐ Factor Xa ☐ SUMO ☐ Thrombin ☐ TEV

☐ **Fig. 4.3** Decision making in selecting a fit-for-purpose plasmid.

4.2.4.1 **Promoter**

Promoters are broadly divided into two main types:

- Constitutive promoters – These are unregulated promoters that allow for continual transcription of their downstream genes.
- Inducible promoters – These promoters are regulated by the presence and the concentration of an abiotic or a biotic factor.

Generally, inducible promoters (*e.g.*, T7*lac*, *ara*BAD) are more frequently used in protein engineering. We have listed some of the most common *E. coli* promoters in ◘ Table 4.7.

The *lac* and T7 are two of the most established promoter systems as you would have deduced from ◘ Tables 4.6 and 4.7. The concepts of these two systems are illustrated in ◘ Figs. 4.4 and 4.5 respectively. The *lac* promoter is a constitutive promoter when used alone. Addition of the *lac* operator sequence enables the LacI repressor to bind, thereby creating an inducible system (◘ Fig. 4.4). The LacI repressor blocks the RNA polymerase from transcribing the target gene, resulting in an "Expression OFF" state. When an inducer (*e.g.*, IPTG) is added, it binds to the LacI repressor, causing it to lose its binding to the *lac* operator. This allows the RNA to transcribe the target gene, leading to an "Expression ON" state. The T7 promoter is constitutive and works specifically with the T7 polymerase (◘ Fig. 4.5). Expression host (λDE3 lysogens) produce T7 polymerase under the control of a *lac* promoter derivative plus the *lac* operator sequence. This creates an inducible system where protein is expressed only when IPTG is added to induce T7 polymerase production, which is required to transcribe the target gene.

There are two important considerations when choosing a promoter: leakiness and protein toxicity (◘ Table 4.8). These are common problems in recombinant protein expression in *E. coli*.

4.2.4.2 **Protein Tag**

Protein tagging is an important aspect of protein science and protein engineering. Fusion tags can be a few amino acids, polypeptides, small proteins or enzymes added to the amino (N) or carboxy (C) terminus of a protein. Tagging is most commonly done via cloning the gene of interest into vectors with tags already incorporated (see ◘ Table 4.6).

◘ Table 4.9 provides an overview of commonly used protein tags. It is worth mentioning that some protein tags offer multiple benefits. For example, GST, MBP and SUMO tags are used mainly for protein purification, but they are also widely known to improve expression yield and protein solubility.

4.2.4.3 **Protease Cleavage**

A protein tag can be cleaved off using a protease to remove any potential interference from the protein tag. Frequently used proteases are given in ◘ Table 4.10. If you

◻ Table 4.7 *E. coli* promoters

Promoter	Description	Expression
*ara*BAD	Promoter of the arabinose metabolic operon	Inducible by arabinose.
lac	Promoter from the *lac* operon	Constitutive. Can be made IPTG-inducible by addition of the *lac* operator sequence.
*lac*UV5	Mutated promoter from the *lac* operon	Regulated like the *lac* promoter.
phoA	Promoter of the bacterial alkaline phosphatase gene	Inducible by phosphate starvation.
pL	Promoter from λ bacteriophage	Regulated by temperature.
pR	Promoter from λ bacteriophage	Regulated by temperature.
*rha*BAD	Promoter of the *rhaBAD* operon	Inducible by rhamnose.
Sp6	Promoter from Sp6 bacteriophage	Constitutive, but requires SP6 RNA polymerase.
T3	Promoter from T3 bacteriophage	Constitutive, but requires T3 RNA polymerase.
T5	Promoter from T5 bacteriophage	Constitutive. Can be made IPTG-inducible by addition of the *lac* operator sequence.
T7	Promoter from T7 bacteriophage	Constitutive, but requires T7 RNA polymerase. Can be made inducible by production of the T7 RNA polymerase in expression hosts (λDE3 lysogens) under the control of a *lac* promoter derivative, the *E. coli* L8-UV5 *lac* promoter.
T7*lac*	Promoter from T7 bacteriophage plus the *lac* operators	Inducible. Lower basal expression compared to T7 when not induced. Requires T7 RNA polymerase.
tac	Hybrid promoter of *lac* and *trp*	Regulated like the *lac* promoter
tet	Promoter/operator region from the *tet*A resistance gene	Inducible by anhydrotetracycline.
trc	Hybrid promoter of *lac* and *trp*	Regulated like the *lac* promoter
trp	Promoter from the *E. coli* tryptophan operon	Inducible. Repressed by Trp and induced by causing a Trp deficiency with indole-2-acrylic acid

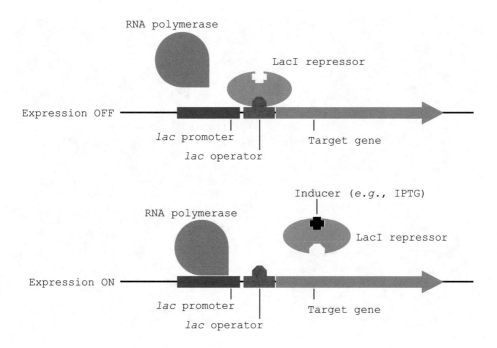

Fig. 4.4 The *lac* promoter system.

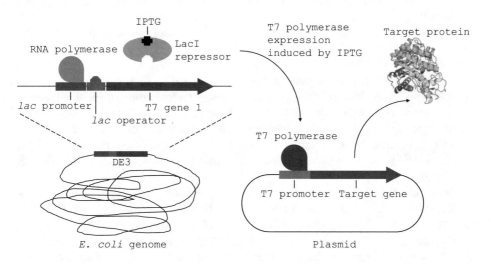

Fig. 4.5 The T7 promoter system.

4

□ Table 4.8   Key considerations when choosing a promoter

Key considerations	Description	Solutions
Leakiness	Some promoters (*e.g.*, *lac* and T7) are not tightly regulated and show basal expression even without the addition of an inducer. Leakiness could lead to plasmid instability and/ or plasmid loss.	• Use a more tightly regulated promoter such as *ara*BAD. • Use an expression host harbouring a compatible pLysS or pLysE plasmid. These plasmids carry the gene encoding T7 lysozyme. T7 lysozyme reduces the basal expression of target genes by inhibiting T7 RNA polymerase. But it does not interfere with the level of expression achieved following the induction by IPTG. • Add 1% (w/v) glucose to the culture medium to repress the induction of *lac* promoter by lactose.
Protein toxicity	Some proteins (*e.g.*, proteases) are toxic to the bacterial cells. Their expression inhibits bacterial growth.	• Suppress promoter leakiness as discussed above. • Use a lower protein expression temperature. • Use a promoter of lower strength.

want to bypass the use of a separate protease, which adds to the protein production cost, you could consider using self-cleaving protein elements such as:

— Modified *Saccharomyces cerevisiae* vacuolar ATPase subunit A intein (*Sce* VMA intein)
— ΔI-CM mini-intein developed from the *Mycobacterium tuberculosis* RecA intein

4.2.4.4 Co-Plasmid

Beside leaky expression and protein toxicity discussed in □ Table 4.8, other common protein expression problems in *E. coli* include the formation of inclusion bodies and protease degradation of the protein. These problems are often a consequence of improper folding of the expressed proteins. A potential solution is to use a co-plasmid that expresses a molecular chaperone or a group of chaperones to facilitate protein folding. Commonly used chaperones include DnaK, DnaJ, GrpE, GroES, GroEL and Tig.

When using a co-plasmid, it is important to pay attention to the plasmid incompatibility (□ Table 4.11). Generally, if two plasmids are members of the same incompatibility group, the introduction of one of the two plasmids by conjugation, transformation, or transduction into a cell carrying the other plasmid destabilizes the inheritance of the resident plasmid. This is because plasmids with the same origin of replication will compete for the same machinery for plasmid replication.

■ **Table 4.9** Commonly used protein tags, their sizes and applications

Application	Tag	Size[a] [kDa]
Affinity protein purification	Biotin-carboxy carrier protein (BCCP)	17
	Calmodulin	17
	Chitin-binding domain (CBD)	6
	Glutathione-S-transferase (GST)	26
	HaloTag	34
	Maltose-binding protein (MBP)	42
	Polyhistidine (6×His; `HHHHHH`)	1
	SBP-tag (`MDEKTTGWRGGHVVEGLAGELEQLRARLEHH-PQGQREP`)	4
	Strep-tag II (`WSHPQFEK`)	1
	Twin-Strep-tag [`WSHPQFEK(GGGS)`$_2$`GGSAWSHPQFEK`]	3
Epitope tag (suitable for Western blot, immunoprecipitation, immunochemistry, and affinity purification)	FLAG tag (`DYKDDDDK`)	1
	HA tag (`YPYDVPDYA`)	1
	Myc tag (`EQKLISEEDL`)	1
	S tag (`KETAAAKFERQHMDS`)	2
	Softtag 3 (`TQDPSRVG`)	1
	T7 tag (`MASMTGGQQMG`)	1
	V5 tag (`GKPIPNPLLGLDST`)	1
	Xpress tag (`DLYDDDDK`)	1
Protein expression yield and solubility	AFV$_{1-99}$ from *Acidianus* filamentous virus (AFV)	14
	Aggregation-resistant protein (SlyD)	22
	AmpC-type β-lactamase (Bla)	40
	Disulphide isomerase I (DsbA)	21
	Elongation factor Ts (Tsf)	31
	Fasciola hepatica antigen (Fh8)	8
	Lipoyl domain from *Bacillus stearothermophilus* E2p	12
	N-utilization substance A (NusA)	55
	Peptidyl-prolyl *cis–trans* isomerase B (PpiB)	18
	Thioredoxin (Trx)	12
Labelling/ detection	Acyl carrier protein (ACP) tag	9
	AviTag (`GLNDIFEAQKIEWHE`)	2
	CLIP tag	20
	Escherichia coli dihydrofolate reductase (eDHFR)	18
	FlAsH/ReAsH tag (`CCGPCC`)	1
	Green fluorescent protein (GFP)	27

(continued)

◘ Table 4.9 (continued)

Application	Tag	Size^a [kDa]
	HiBiT (`VSGWRLFKKIS`)	1
	Mutant acyl carrier protein (MCP) tag	9
	SNAP tag	20
	SpyTag (`AHIVMVDAYKPTK`)	1
Protein secretion	Alkaline phosphatase of *E. coli* (PhoA; `MKQSTIALALLPLLFTPVTKA`)	2
	Cell division protein of *E. coli* (SufI; `MSLSRRQFIQASGIALCAGAVPLKASA`)	3
	D-galactose-binding periplasmic protein of *E. coli* (MglB; `MNKKVLTLSAVMASMLFGAAAHA`)	2
	Endo-1,4- β-xylanase from *Bacillus velezensis* (EOX; `MFKFKKKFLVGLTAAFMSISMFSATASA`)	3
	Fimbrial chaperone of *E. coli* (SfmC; `MMTKIKLLMLIIFYLIISASAHA`)	3
	Heat-stable enterotoxin II of *E. coli* (STII; `MKKNIAFLLASMFVFSIATNAYA`)	3
	Maltoporin of *E. coli* (LamB; `MMITLRKLPLAVAVAAGVMSAQAMA`)	3
	Maltose/maltodextrin-binding periplasmic protein of *E. coli* (MalE; `MKIKTGARILALSALTTMMFSASALA`)	3
	Minor capsid protein from fd phage (gIII; `MKKLLFAIPLVVPFYSHS`)	2
	Outer membrane protein A of *E. coli* (OmpA; `MKKTAIAIAVALAGFATVAQA`)	2
	Outer membrane protein C of *E. coli* (OmpC; `MKVKVLSLLVPALLVAGAANA`)	2
	Outer membrane protein T of *E. coli* (OmpT; `MRAKLLGIVLTTPIAISSFA`)	2
	Osmotically-inducible protein Y of *E. coli* (OsmY)	21
	Pectate lyase B of *Erwinia carotovora* (PelB; `MFKYLTPIFLCTAAFSFQAQA`)	2
	Periplasmic protein of *E. coli* (TorT; `MRVLLFLLLSLFMLPAFS`)	2
	Phosphatase PAP2 family protein of *E. coli* (MmAp; `MKKNIIAGCLFSLFSLSALA`)	2
	Thiol:disulfide interchange protein of *E. coli* (DsbA; `MKKIWLALAGLVLAFSASA`)	2
	Tol-Pal system protein of *E. coli* (TolB; `MKQALRVAFGFLILWASVLHA`)	2
	YebF of *E. coli*	13

^aProtein size rounded to the nearest kDa

□ Table 4.10 List of proteolytic cleavage sites commonly found in bacterial plasmids

Protease	Protease cleavage site	Scar after cleavage
Enterokinase	DDDDK↑	✗
Human rhinovirus 3C (HRV-3C) protease	LEVLFQ↑GP	✓
Factor Xa protease	I(E/D)GR↑	✗
SUMO protease	Recognizing the tertiary structure of SUMO rather than an amino acid sequence	✗
Thrombin	LVPR↑GS	✓
Tobacco etch virus (TEV) protease	ENLYFQ↑(G/S)	✓

□ Table 4.11 Plasmid incompatibility

Origin of replication	Copy number	Incompatibility (Inc) group	Examples
ColE1	15–20	A	pColE1
ColE1 derivative and F1	300–500	A	pBluescript
p15A	10	B	pACYC
pBR322	15–20	A	pET, pGEX
pMB1	15–20	A	pBR322
pMB1 derivative	500–700	A	pUC
pSC101	5	C	pSC101
pUC and F1	300–500	A	pGEM
R6K	15–20	C	pR6K

4.3 Choosing an Appropriate Host

Now that you have developed a good understanding of *E. coli* strains and plasmids, let's conclude this chapter by summarizing some useful tips on selecting the right host for your protein expression (□ Table 4.12).

Exercise

(a) What is the antibiotic selection marker in this plasmid? (□ Fig. 4.6)
(b) What is the promoter used for protein expression?

4

Take-Home Messages

1. A variety of expression platforms have been developed for recombinant protein production, including bacteria, yeasts, insect cells, mammalian cells, and fungi *etc.*
2. *E. coli* is the most widely used protein expression system in protein engineering.
3. The Coli Genetic Stock Center (CGSC) contains a huge collection of non-pathogenic laboratory strains of *E. coli*, which are widely used in genetic and molecular studies.
4. DH5α, TOP10F' and XL1-Blue are examples of *E. coli* strains mainly used for routine cloning or subcloning.
5. BL21, BL21(DE3) and C41(DE3) are examples of *E. coli* strains mainly used for recombinant protein expression.
6. Addgene is a non-profit global plasmid repository that archives and distributes plasmids for scientists.
7. SnapGene Viewer is an excellent tool to read a plasmid DNA sequence and to create a plasmid map for visualization.
8. A wide range of *E. coli* plasmids are available. Variations among these plasmids are typically found in the promoter, MCS, tag, tag position, tag cleavage site, selection marker, and the number of expression cassettes.
9. Promoters are broadly divided into two main categories, constitutive promoters and inducible promoters.
10. Protein tags are used for a variety of purposes, including protein purification, protein tracking, improving protein expression yield & solubility, protein labelling and detection, and protein secretion.
11. A protein tag can be removed via a proteolytic cleavage using a protease.
12. For a successful recombinant protein production, it is important to find the right plasmid and expression host combination.

◻ **Table 4.12**   Finding the right *E. coli* host

Situation		*E. coli* strains
Promoter	Use of T7 promoter for protein expression	DE3 strains
	Leaky T7 promoter	Strains harbouring pLysS, pLysE or pLacI plasmid
Target gene	Target gene contains rare codons, resulting in poor protein expression	CodonPlus, Rosetta or strains harbouring pRARE/pRARE2 plasmid
Target protein	Expressing a membrane protein	C41, C43 or Lemo21
	Expressing and purifying a polyhistidine tagged protein	NiCo21 strains
	Expressing a toxic protein	BL21-AI or BL21-Gold
	Promote disulphide bond formation in cytoplasm	Origami, SHuffle or AD494 strains
Protein quality	Avoid endotoxin from *E. coli*	ClearColi

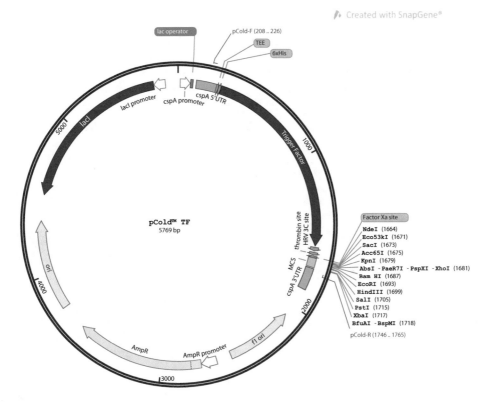

☐ **Fig. 4.6** Plasmid map of pCOLD TF, created using SnapGene.

(c) What is the protein sequence of the translation enhancing element (TEE) used in this plasmid?
(d) What is the affinity tag used for protein purification?
(e) What is the function of a trigger factor?
(f) How large is the trigger factor in kDa?
(g) Which protease would you use to remove the protein fusion tag?
(h) Would you use NdeI and SacII to clone a target gene into this vector?
(i) Name an *E. coli* strain for protein expression from this plasmid.

Further Reading

Baneyx F (1999) Recombinant protein expression in *Escherichia coli*. Curr Opin Biotechnol 10(5): 411–421
Esposito D, Chatterjee DK (2006) Enhancement of soluble protein expression through the use of fusion tags. Curr Opin Biotechnol 17(4):353–358
Kimple ME, Brill AL, Pasker RL (2013) Overview of affinity tags for protein purification. Curr Protoc Protein Sci 73:9 9 1–9 9 23
Young CL, Britton ZT, Robinson AS (2012) Recombinant protein expression and purification: a comprehensive review of affinity tags and microbial applications. Biotechnol J 7(5):620–634

Gene Cloning

Contents

© Springer Nature Switzerland AG 2020
T. S. Wong, K. L. Tee, *A Practical Guide to Protein Engineering*, Learning Materials in Biosciences,
https://doi.org/10.1007/978-3-030-56898-6_5

What You Will Learn in This Chapter

In this chapter, we will learn to:

- search and order a bacterial strain or its genomic DNA from a culture collection
- cultivate a bacterial strain and extract its genomic DNA
- quantify DNA concentration and assess its purity
- design and order oligonucleotides
- perform gene cloning using classical restriction and ligation method
- perform gene cloning using NEBuilder® HiFi DNA Assembly
- perform gene cloning using PTO-QuickStep
- verify a clone with DNA sequencing
- design DNA sequencing primer

Gene cloning is the starting point for protein engineering. The best way to learn gene cloning is to do it and this is the approach we will take in this chapter. In ▶ Chap. 2, we have analysed the sequence of Tfu_0901 extensively. We will now use this endo-β-1,4-glucanase from *T. fusca* YX as our target gene [Tfu_0901 (UniProt Q47RH8, GenBank AAZ54939.1)], and clone it into the pET-19b vector (◨ Fig. 5.1). The MCS of this plasmid is shown in ◨ Fig. 5.2. We will clone this gene using two different starting material:

- Amplifying the Tfu_0901 gene from *T. fusca* YX genomic DNA (gDNA)
- Ordering a synthetic gene of Tfu_0901

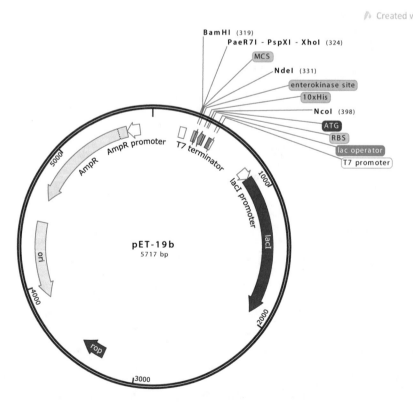

◨ **Fig. 5.1**   Plasmid map of pET-19b, created using SnapGene.

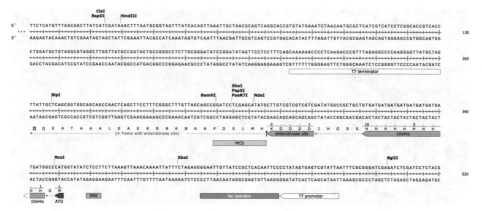

□ **Fig. 5.2** The multiple cloning site of pET-19b, created using SnapGene.

5.1 Gene Amplification from Genomic DNA

Gene amplification from gDNA involves three basic steps:
1. Order *T. fusca* YX strain or its gDNA
2. Cultivate the bacterium and extract its gDNA (if you are purchasing gDNA directly in step 1, you can skip step 2)
3. Amplify Tfu_0901 gene in a PCR

5.1.1 Ordering a Bacterial Strain or Its Genomic DNA

There are a few excellent culture collections, from which you can purchase or request for a strain (□ Table 5.1). To give you a quick guide, let's order *T. fusca* YX from ATCC:
— Go to the ATCC webpage (▶ https://www.lgcstandards-atcc.org/)
— Type `Thermobifida fusca YX` in the query box
— Click the 'Search' button
— To order the bacterial strain, choose 'ATCC® BAA-629', enter the quantity, and 'Add to Cart' (the bacterium will be delivered in a freeze-dried format)
— To order the gDNA, choose 'ATCC® BAA-629D-5', enter the quantity, and 'Add to Cart'

Most culture collections, if not all, provide useful information about the strain. If you click on the link corresponding to ATCC® BAA-629 (▶ https://www.lgcstandards-atcc.org/products/all/BAA-629.aspx), you will gather the following information:
— Under the 'General Information' tab, this is a biosafety level 1 organism.
— Under the 'Culture Method' tab, this aerobic organism grows at 50 °C. The recommended growth media include TYG medium (3 g/L tryptone, 3 g/L yeast extract, 3 g/L glucose, 1 g/L K_2HPO_4, pH 7.4) and Hagerdahl medium.
— Under the 'History' tab, relevant literatures and genome information (GenBank CP000088.1) are listed.
— Under the 'Documentation' tab, you can download the product sheet and the safety data sheet.

▢ Table 5.1   Culture collections

Culture collection	URL	Fee?
Agricultural Research Service Culture Collection (NRRL)	▶ https://nrrl.ncaur.usda.gov/	No
American Type Culture Collection (ATCC)	▶ https://www.lgcstandards-atcc.org/	Yes
German Collection of Microorganisms and Cell Cultures GmbH (DSMZ)	▶ https://www.dsmz.de/	Yes
Japan Collection of Microorganisms (JCM)	▶ https://jcm.brc.riken.jp/en/	Yes
National Collection of Industrial Food and Marine Bacteria (NCIMB)	▶ https://www.ncimb.com/	Yes
Thailand Bioresource Research Center (TBRC)	▶ https://www.tbrcnetwork.org/	Yes

▢ Table 5.2   gDNA extraction kits

Supplier	Kit
Macherey-Nagel	NucleoSpin Microbial DNA Mini kit
Qiagen	QIAamp DNA Mini Kit
	Genomic-tip 20/G
Thermo Fisher Scientific	GeneJET Genomic DNA Purification Kit
	PureLink™ Genomic DNA Mini Kit
Omega Bio-tek	E.Z.N.A.® Bacterial DNA Kit
New England Biolabs (NEB)	Monarch® Genomic DNA Purification Kit

5.1.2 Bacterial Cultivation and Genomic DNA Extraction

Upon receiving the strain from ATCC, we would advise that you propagate the strain immediately by following the propagation procedure recommended in the product sheet. For long-term storage, we recommend creating a few tubes of bacterial glycerol stock:

- Inoculate an overnight liquid culture.
- Once you have a dense culture, mix 500 µL of the overnight culture and 500 µL of 50% (v/v) glycerol in a 2-mL cryogenic tube.
- Freeze the glycerol stock tube at −80 °C.

Once you have prepared your glycerol stock, you can proceed to gDNA extraction. There are many commercial kits developed for this purpose, as indicated in ▢ Table 5.2. In our experience, there isn't a significant difference in the performance between these kits, as long as you follow the manufacturer's instructions. In fact, most of the kits follow the steps below:

1. Preparation of cell lysate with a lysis buffer, which typically contains Triton X-100, Tween-20, guanidinium thiocyanate, and/or guanidine hydrochloride. In some cases, lysozyme/lysostaphin is added or glass beads are used to facilitate cell lysis. Lysis buffer is also supplemented with RNase to remove RNA.
2. Protein digestion with a protease (typically Proteinase K).
3. DNA binding to a silica membrane or anion-exchange resin.
4. DNA washing with buffer containing ethanol.
5. DNA elution in TE buffer (10 mM Tris-HCl, 0.1 mM EDTA, pH 8.0).

The availability of commercial kits has simplified our experimental work tremendously. However, it remains important to understand the principles behind the protocol. Especially when using the kit for the first time, our advice is:

— Read the accompanying manual as it usually contains ample useful information (*e.g.*, principle of the kit, protocols, buffer composition, and troubleshooting guide *etc*) before you start.
— Follow the protocol strictly the first time you use it.
— As you become more experienced, you may adjust it as appropriate for your sample.

Upon gDNA extraction, DNA yield is determined from the DNA concentration in the eluate, measured by absorbance at 260 nm with a NanoDrop or a VersaWave. Purity is determined by calculating the ratio of absorbance at 260 nm to absorbance at 280 nm and the ratio of absorbance at 260 nm to absorbance at 230 nm (◘ Table 5.3). Usually, A_{260}/A_{280} and A_{260}/A_{230} ratios of >1.8 indicate a 'clean' prep, suitable for most downstream applications such as PCR and DNA sequencing.

◘ **Table 5.3** Assessing nucleic acid purity

Parameter	Explanation	Acceptable range for pure dsDNA or RNA
A_{260}/A_{280}	This ratio provides an insight regarding the type of nucleic acid present (dsDNA or RNA) as well as a rough indication of purity. Typically, protein contamination can be detected by a reduction of this ratio; RNA contamination can be detected by an increase of this ratio.	• dsDNA: 1.85–1.88 • RNA: ~2.1
A_{260}/A_{230}	This ratio is an indicator of contaminants that absorb at 230 nm, including chaotropic salts such as guanidine thiocyanate and guanidine hydrochloride, EDTA, non-ionic detergents like Triton X-100 and Tween® 20, and phenol. Substances like polysaccharides or free floating solid particles like silica fibres absorb at this wavelength, but will show a weaker effect.	• dsDNA: 2.30–2.40 • RNA: 2.10–2.30

5.1.3 Gene Amplification

To amplify Tfu_0901 gene, you would need to follow the procedure below:
- Have the DNA sequence of Tfu_0901 ready (▶ Sect. 2.1)
- Have the plasmid map ready (▶ Sect. 4.2)
- Decide on your cloning strategy after checking the gene sequence and the plasmid map
- Design your primers accordingly
- Order your primers
- Conduct a PCR
- Run a DNA gel to check your PCR product

5.1.3.1 Primer Design and Ordering Primers

Primer design is governed by your cloning strategy, which requires prior analysis on the target gene and careful selection of the right plasmid. Using the cloning of Tfu_0901 into pET-19b as an example, here are our considerations before designing primers:
- Looking at the plasmid map (◘ Fig. 5.1) and the MCS (◘ Fig. 5.2) of pET-19b, we could potentially use NdeI and BamHI sites to clone Tfu_0901. These are unique restriction sites, highlighted in bold in both the SnapGene plasmid map and the MCS sequence. Using NdeI site will ensure that the gene is in frame with the enterokinase site.
- Tfu_0901 contains a signal peptide (amino acids 1–36), for the purpose of secretion by *T. fusca* YX. Since we are looking to express this gene in *E. coli*, we would need to exclude this stretch of sequence in our gene amplification.
- Next, we use either SnapGene or NEBCutter (▶ http://nc2.neb.com/NEBcutter2/) to confirm that neither NdeI nor BamHI is present in the Tfu_0901 gene.
- Therefore, our forward primer will contain NdeI site, and the reverse primer will contain a stop codon and BamHI site. *E. coli* displays a strong bias towards the TAA stop codon, and translational termination efficiency is further improved by the prolonged TAAT sequence. Alternatively, two consecutive stop codons ensure termination.

Primer design is straightforward. Good primers usually display the following properties:
- Length of 18–24 bases
- GC content of 40–60%
- End with 1 or 2 G or C, but not more than 2
- Melting temperature (T_m) of 50–60 °C
- The primer pairs should have a T_m value within 5 °C of each other
- The primers should not have complementary regions

Depending on the restriction sites you choose and the DNA sequence of your target gene, it is not always possible to satisfy all the above requirements. However, these recommendations are not strict requirements and it remains possible to amplify your target gene.

Based on the cloning strategy and primer design considerations, we have designed the primers as shown in ◘ Table 5.4 and ◘ Fig. 5.3 to amplify Tfu_0901$_{37-466}$. Each

◼ Table 5.4   Primers for Tfu_0901 amplification

Fwd primer

5'–TATACATATGGCCGGTCTCACCGCCACAG–3'

Colour	Sequence (5'→3')	Function
	TATA	Additional bases for more effective restrictive digestion
	CATATG	NdeI site (already incorporated start codon, ATG)
	GCCGGTCTCACCGCCACAG	Region annealed to the template DNA

Rev primer

5'–TATAGGATCCTTAGGACTGGAGCTTGCTCCGCAC–3'

Colour	Sequence (5'→3')	Function
	TATA	Additional bases for more effective restrictive digestion
	GGATCC	BamHI site
	TTA	Stop codon
	GGACTGGAGCTTGCTCCGCAC	Region annealed to the template DNA

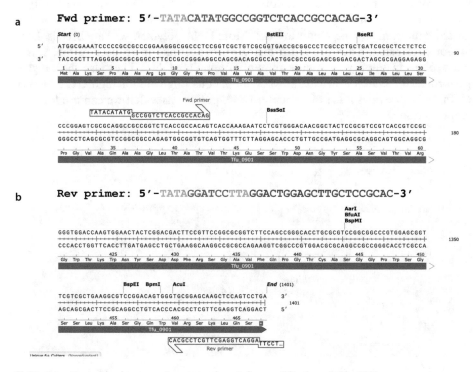

◼ Fig. 5.3   Forward primer **a** and reverse primer **b** for amplification of Tfu_0901.

primer is composed of multiple elements, which are all essential. Important to note, you do have the flexibility in changing the sequence of the additional bases (coloured in orange) to avoid hairpin formation and self-complementarity. OligoCalc (▶ http://biotools.nubic.northwestern.edu/OligoCalc.html) is very useful for calculating oligonucleotide properties and checking any potential self-complementarity. You can also change the length of the region that anneals to the template DNA (coloured in blue) to adjust the primer T_m value (which is influenced by the GC content of your primer).

Now that we have our primers, let's order them! There are many oligonucleotide suppliers (*e.g.*, Eurofins and Integrated DNA Technologies). When placing an order, the primer sequence is always reported in 5′→3′. We also suggest the following to reduce cost:

= For unmodified oligos, choose the smallest synthesis scale (0.01 μmol).
= Go for salt-free oligos.

Oligos are usually delivered in a lyophilised format. We advise resuspending the DNA in water to a concentration of 100 μM (stock solution) and dilute to a working solution of 10 μM or 20 μM (again in water). For long-term storage, place the oligo solution in −20 °C freezer.

5.1.3.2 Polymerase Chain Reaction (PCR) and DNA Polymerases

PCR is a routine method in protein engineering laboratories. Kary B. Mullis (1944–2019), an American biochemist, is credited with inventing the technique in 1983 while working for Cetus Corporation in Emeryville, California. He won the 1993 Nobel Prize in Chemistry for his invention. Over the past few decades, we have witnessed significant advancements in PCR. More DNA polymerases have been isolated and/or engineered. They are much faster, more accurate in replicating DNA, and more robust in amplifying minute amount of template DNA. They also show higher performance towards a broad range of amplicons (high AT or high GC content). ◘ Table 5.5 provides a list of common DNA polymerases that we can use for gene amplification.

◘ **Table 5.5**   DNA polymerases and their properties, arranged in the order of increasing fidelity

DNA polymerases	Extension rate	3′→5′ exonuclease activity	Fidelity (*vs* Taq)
Taq	1 min/kb	✗	1×
OneTaq	1 min/kb	✓	2×
Platinum Taq HiFi	1 min/kb	✓	6×
KOD	7–20 sec/kb	✓	12×
PfuUltra	1 min/kb	✓	19×
PfuUltra II Fusion HS	15 sec/kb	✓	20×
AccuPrime Pfx	1 min/kb	✓	26×
Phusion	15–30 sec/kb	✓	39–50×
Q5	20–30 sec/kb	✓	280×
Platinum SuperFi II	15–30 sec/kb	✓	300×

Fast and accurate DNA polymerases reduce experimental time and errors during amplification. We have chosen Platinum SuperFi II, Q5 and PfuUltra II Fusion FS to demonstrate the design and setup of a PCR mixture and a thermocycler programme (☐ Tables 5.6 and 5.7). Before we can do that, we need three pieces of information:

- The regions annealing to the template DNA have sequences of 'GCCGGTCTCACCGCCACAG' and 'GGACTGGAGCTTGCTCCGCAC' → to determine annealing temperature.
- The expected amplicon size is 1290 bp (excluding the first 36 amino acids) → to calculate the extension time.
- The GC content of the amplicon is 67% → to decide which buffer to use.

Whilst preparing the PCR mixture, do follow the order presented in ☐ Table 5.6, *i.e.*, water first, buffer next, enzyme last! As you are pipetting small volumes of liquid, do make sure the pipette tip goes into water before dispensing. Last but not least, mix well before you start your PCR.

After PCR, an agarose gel electrophoresis is used to analyse the DNA product. We summarise four potential PCR outcomes in ☐ Table 5.8.

☐ **Table 5.6** Preparing PCR mixture

Platinum SuperFi II		Q5		PfuUltra II Fusion FS	
Component	**Volume [μL]**	**Component**	**Volume [μL]**	**Component**	**Volume [μL]**
Water	36 − X	Water	26.5 − Y	Water	41 − Z
5× SuperFi II buffer	10	5× Q5 reaction buffer	10	10×PfuUltra II reaction buffer	5
		5× Q5 high GC enhancer	10		
10 mM dNTP	1	10 mM dNTP	1	10 mM dNTP	1
20 μM Fwd primer[a]	1	20 μM Fwd primer[a]	1	20 μM Fwd primer[a]	1
20 μM Rev primer[b]	1	20 μM Rev primer[b]	1	20 μM Rev primer[b]	1
gDNA	X (100 ng)	gDNA	Y (100 ng)	gDNA	Z (100 ng)
Platinum SuperFi II DNA polymerase	1	2 U/μL Q5 high-fidelity DNA polymerase	0.5	PfuUltra II Fusion HS DNA polymerase	1
Total	**50**	**Total**	**50**	**Total**	**50**

[a]Use the Fwd primer in ☐ Table 5.4
[b]Use the Rev primer in ☐ Table 5.4

□ Table 5.7 Thermal cycling conditions

Platinum SuperFi II			Q5			PfuUltra II Fusion FS		
Temperature [°C]	Duration	# of cycle	Temperature [°C]	Duration	# of cycle	Temperature [°C]	Duration	# of cycle
98	30 sec	1×	98	30 sec	1×	95	2 min	1×
98	8 sec	30×	98	8 sec	30×	95	20 sec	30×
60[a]	10 sec		72[c]	20 sec		60[e]	20 sec	
72	29 sec[b]		72	33 sec[d]		72	20 sec[f]	
72	5 min	1×	72	2 min	1×	72	3 min	1×
8	∞	–	8	∞	–	8	∞	–

[a]The annealing temperature with Platinum™ SuperFi™ II DNA Polymerase is 60 °C. Proprietary additives in the reaction buffer stabilize primer-template duplexes during the annealing step, and contribute to increased specificity without the need to optimize annealing temperature for each primer pair
[b]Calculated using 1290 bp/1000 bp × ((15 sec + 30 sec)/2) = 29 sec
[c]Calculated using NEBTm calculator (▶ https://tmcalculator.neb.com/#!/main) with sequences 'GCCGGTCTCACCGCCACAG' and 'GGACTGGAGCTTGCTCCGCAC'
[d]Calculated using 1290 bp/1000 bp × ((20 sec + 30 sec)/2) = 32.3 sec ≈ 33 sec
[e]Calculated using T_m – 5 °C = 65 °C – 5 °C = 60 °C (T_m of 65 °C is determined using SnapGene)
[f]Calculated using 1290 bp/1000 bp × 15 sec = 19.4 sec ≈ 20 sec

□ Table 5.8 Outcomes of a PCR

Outcome	Product of right size	Side bands or by-products	What next?
1	✓	✗	• Add 20 U of DpnI directly into the PCR mixture, and incubate at 37 °C for 2 hours to degrade gDNA • Purify the PCR product using a DNA clean-up kit
2	✓	✓	• Purify the PCR product by gel extraction
3	✗	✓	• Repeat with another DNA polymerase
4	✗	✗	• Repeat the PCR with more care • Check that the primers are right

5.2 Gene Synthesis

Compared to gene amplification from gDNA (▶ Sect. 5.1), gene synthesis is a more 'hands-off' approach. Some might argue that it is too expensive. However, with the rapid advancement in DNA technologies, the price of a synthetic gene is now very competitive. Let's do a cost comparison to convince ourselves (□ Table 5.9).

□ Table 5.9 Cost comparison between gene amplification from genomic DNA and gene synthesis

Starting material	Items to purchase	Unit price [£]	Total cost [£]
Bacterial strain (► Sect. 5.1)	Bacterial strain gDNA extraction kit	• 390 (for ATCC® BAA-629) • 169 [for QIAamp DNA Mini Kit (50)]	559.00
Genomic DNA (► Sect. 5.1)	Genomic DNA	• 259 (ATCC® BAA-629D-5)	259.00
Synthetic gene (► Sect. 5.2)	Synthetic gene	• 245 (£0.17 per base)	219.30

□ Fig. 5.4 Tfu_0901 gene design for synthesis.

5.2.1 **Gene Design**

Using a synthetic gene has multiple benefits:

— Hassle-free
— Control over gene design (you can design the gene elements such as tag, cleavage site, and restriction sites *etc* in precisely the way you want them)
— Codon optimization to improve protein expression yield
— Create gene sequences that do not exist in Nature
— Introduce desired mutation(s)
— Remove unwanted restriction sites
— Verified gene sequence

The synthetic gene should be designed to match the plasmid and cloning strategy we want to use. To clone Tfu_0901$_{37-466}$ into pET-19b (□ Fig. 5.1) using NdeI and BamHI sites (□ Fig. 5.2), we designed the sequence in □ Fig. 5.4. In this design, the protein sequence (in capital letters) is flanked by two nucleotide sequences (in small letters). The flaking nucleotide sequences allow for incorporation of restriction sites and stop codon. The protein sequence will be codon-optimized for expression in *E. coli* (the chosen expression host).

5.2.2 Ordering a Synthetic Gene and Codon Optimization

As mentioned briefly in ▶ Sect. 2.2.2.5, a host of companies offer gene synthesis service such as Eurofins, Integrated DNA Technologies, GenScript, and GeneWiz *etc.* Many of them have their in-house gene optimization algorithm. When placing your order online, you need to:

- Provide the protein sequence you want to codon-optimize
- Specify which organism you are optimizing the codons for
- Insert 5′- and 3′-flaking sequences
- Specify the restriction sites you would like to avoid within the optimised gene sequence to enable subsequent cloning

Let's optimize $Tfu\_0901_{37\text{-}466}$ using algorithms from various service providers. The sequences optimized by Eurofins (GENEnius), GeneWiz and GenScript (GenSmart) show similar GC content: 54.65% (Eurofins), 59.69% (GeneWiz), and GenScript (56.28%). ◻ Figure 5.5 compares the codon usage of the optimized sequences. It is

AA	Codon	E. coli K12	E. coli B	Eurofins Fraction	GeneWiz Fraction	GenScript Fraction
Ala	GCG	0.38	0.36	0.37	0.49	0.46
Ala	GCA	0.21	0.21	0.20	0	0.17
Ala	GCT	0.11	0.15	0.17	0	0.26
Ala	GCC	0.31	0.28	0.26	0.51	0.11
Cys	TGT	0.42	0.42	0.50	0.33	0.33
Cys	TGC	0.58	0.58	0.50	0.67	0.67
Asp	GAT	0.65	0.66	0.64	0.68	0.44
Asp	GAC	0.35	0.34	0.36	0.32	0.56
Glu	GAG	0.30	0.38	0.33	0.24	0.48
Glu	GAA	0.70	0.62	0.67	0.76	0.52
Phe	TTT	0.57	0.61	0.60	0.13	0.47
Phe	TTC	0.43	0.39	0.40	0.87	0.53
Gly	GGG	0.12	0.18	0.15	0	0
Gly	GGA	0.13	0.10	0.11	0	0.07
Gly	GGT	0.29	0.30	0.33	0.35	0.52
Gly	GGC	0.46	0.41	0.41	0.65	0.41
His	CAT	0.55	0.56	0.58	0.33	0.42
His	CAC	0.45	0.44	0.42	0.67	0.58
Ile	ATA	0.07	0.07	0	0	0
Ile	ATT	0.58	0.48	0.57	0.10	0.43
Ile	ATC	0.35	0.44	0.43	0.90	0.57
Lys	AAG	0.27	0.23	0.20	0.30	0.40
Lys	AAA	0.73	0.77	0.80	0.70	0.60
Leu	TTG	0.12	0.15	0.14	0	0.19
Leu	TTA	0.15	0.14	0.14	0	0.05
Leu	CTG	0.46	0.45	0.52	0.86	0.62
Leu	CTA	0.05	0.03	0	0	0.05
Leu	CTT	0.12	0.11	0.10	0	0.05
Leu	CTC	0.10	0.12	0.10	0.14	0.05
Met	ATG	1.00	1.00	1.00	1.00	1.00
Asn	AAT	0.47	0.57	0.47	0.37	0.37
Asn	AAC	0.53	0.43	0.53	0.63	0.63

AA	Codon	E. coli K12	E. coli B	Eurofins Fraction	GeneWiz Fraction	GenScript Fraction
Pro	CCG	0.55	0.61	0.52	0.52	0.80
Pro	CCA	0.14	0.18	0.20	0.48	0.12
Pro	CCT	0.17	0.14	0.16	0	0.08
Pro	CCC	0.13	0.06	0.12	0	0
Gln	CAG	0.70	0.65	0.63	0.56	0.56
Gln	CAA	0.30	0.35	0.38	0.44	0.44
Arg	AGG	0.03	0.04	0	0	0
Arg	AGA	0.02	0.05	0	0	0.06
Arg	CGG	0.07	0.11	0.11	0	0.06
Arg	CGA	0.07	0.05	0	0	0
Arg	CGT	0.36	0.35	0.39	0.22	0.56
Arg	CGC	0.44	0.40	0.50	0.78	0.33
Ser	AGT	0.14	0.16	0.14	0.29	0.05
Ser	AGC	0.33	0.25	0.29	0.71	0.50
Ser	TCG	0.16	0.20	0.17	0	0.10
Ser	TCA	0.15	0.11	0.12	0	0.05
Ser	TCT	0.11	0.15	0.14	0	0.14
Ser	TCC	0.11	0.14	0.14	0	0.17
Thr	ACG	0.24	0.27	0.25	0.31	0.19
Thr	ACA	0.13	0.11	0.14	0	0.06
Thr	ACT	0.16	0.14	0.17	0	0.17
Thr	ACC	0.47	0.47	0.44	0.69	0.58
Val	GTG	0.40	0.43	0.37	0.37	0.33
Val	GTA	0.17	0.14	0.15	0	0.11
Val	GTT	0.25	0.25	0.26	0.63	0.41
Val	GTC	0.18	0.18	0.22	0	0.15
Trp	TGG	1.00	1.00	1.00	1.00	1.00
Tyr	TAT	0.53	0.69	0.60	0.13	0.47
Tyr	TAC	0.47	0.31	0.40	0.87	0.53
End	TGA	0.36	0.27	0	0	0
End	TAG	0	0.09	0	0	0
End	TAA	0.64	0.64	0	0	0

◻ **Fig. 5.5** Codon usage of the $Tfu\_0901_{37\text{-}466}$ sequences optimized by Eurofins, GeneWiz and GenScript for expression in *E. coli* K12. The restriction sites NdeI and BamHI are avoided in the optimized sequences. The analysis is done using the Codon Usage tool discussed in ▶ Sect. 2.2.2.4. Codon usage of the *E. coli* K12 and B strains from ▶ https://www.kazusa.or.jp/codon/ are included for comparison. Codon for each amino acid with the highest occurrence is coloured in green and with the second highest occurrence in yellow.

clear from this comparison that there are differences between the algorithms. Eurofins's algorithm tends to utilise all the codons for an amino acid, but maximizes the occurrence of the most frequently used codons. In contrast, the GeneWiz algorithm is more 'polarized'. It discounts the less frequently used codons and utilizes only the two most frequently used codons for each amino acid. Which algorithm is better? Well, we do not have an answer to this. Codon optimization does not always guarantee a high protein expression yield. Occasionally, it may even result in a poor protein yield.

A synthetic gene is typically cloned into a standard vector:

- Eurofins: pEX-A128, pEX-A258, pEX-K168 and pEX-K248
- GeneWiz: pUC57-Kan and pUC57-Amp
- GenScript: pUC57-, pUC18- or pUC19-based vectors and a list of expression vectors

When you receive your synthetic gene:

- Transform the plasmid into a cloning strain (*e.g.*, DH5α)
- Prepare a few tubes of bacterial glycerol stock (see ► Sect. 5.1.2) for long-term storage

5.3 Molecular Cloning

Once the gene is amplified, it needs to be cloned into a vector. Molecular cloning methods can be classified into five strategies (◘ Fig. 5.6):

1. Molecular cloning based on complementary overhangs
2. Molecular cloning based on homologous sequences

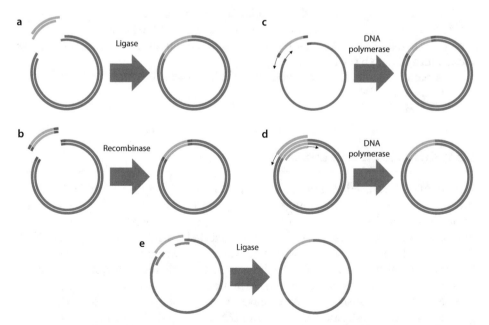

◘ **Fig. 5.6** Cloning strategies: molecular cloning based on complementary overhangs **a**, molecular cloning based on homologous sequences **b**, molecular cloning based on overlapping PCR **c**, molecular cloning based on megaprimers **d**, and molecular cloning based on bridged ligation **e**.

◻ Table 5.10   Commercial enzymes and kits for molecular cloning

Cloning strategy	Enzymes/kits
Complementary overhangs	• TA Cloning • TOPO Cloning • NEBuilder® HiFi DNA Assembly • Gibson Assembly • Golden Gate Assembly • In-Fusion Cloning • USER Cloning
Homologous sequences	• Gateway Cloning • Creator Cloning • Echo Cloning
Overlapping PCR	Not available
Megaprimers	Not available
Bridged ligation	Not available

3. Molecular cloning based on overlapping PCR
4. Molecular cloning based on megaprimers
5. Molecular cloning based on bridged ligation

For some of these strategies, commercial kits are available, as summarised in ◻ Table 5.10.

We are not going to cover all molecular cloning methods in this book. It is more important for students to master one or two techniques at the early stage of learning cloning. Therefore, we have picked three methods to illustrate how Tfu_0901_{37-466} is cloned into pET-19b (◻ Fig. 5.7):

— Restriction and ligation, commonly known as classical cloning
— NEBuilder® HiFi DNA Assembly
— PTO-QuickStep

5.3.1 **Restriction and Ligation**

For cloning by restriction and ligation, two restriction enzymes (double digestion) are used to cut a plasmid (backbone) in order to insert a linear gene fragment (insert) that has been cut by the same set of restriction enzymes. The restriction digests generate complementary overhangs between the digested plasmid and the digested DNA fragment. T4 DNA ligase is subsequently used to covalently join the plasmid and the gene fragment at these sites of complementarity, thereby generating a circular plasmid that is ready for bacterial transformation. There are many enzyme suppliers (*e.g.*, NEB, Thermo Fisher Scientific, and Promega, to name a few), from whom you can purchase your restriction enzymes and DNA ligase.

In ▶ Sects. 5.1 and 5.2 above, we have designed our gene amplification and synthetic gene in a way that enables subsequent gene cloning using restriction and ligation with

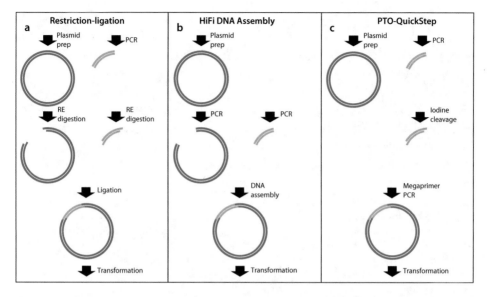

● **Fig. 5.7** Cloning methods: Restriction and ligation **a**, NEBuilder® HiFi DNA Assembly **b**, and PTO-QuickStep **c**.

● **Table 5.11** Plasmid and insert digestion with restriction enzymes from NEB

Plasmid digestion		Insert digestion	
Component	**Volume [μL]**	**Component**	**Volume [μL]**
Water	88 – X	Water	88 – Y
10× CutSmart buffer	10	10× CutSmart buffer	10
pET-19b	X (2 μg)	PCR product or plasmid carrying the synthetic gene of Tfu_0901$_{37-466}$	Y (2 μg)
20 U/μL NdeI	1	20 U/μL NdeI	1
20 U/μL BamHI-HF	1	20 U/μL BamHI-HF	1
Total	100	Total	100

NdeI, BamHI and T4 DNA ligase. In ● Tables 5.11 and 5.12, we will show you how the restriction digestion and ligation using enzymes obtained from NEB can be set up.

Setting up restrictive digestion is straightforward. Typically, we do an overnight digestion at 37 °C. The tips below may prove useful:

- Purify the plasmid backbone after restriction digestion by gel extraction.
- Purify the PCR product after restriction digestion using a DNA clean-up kit or extract the synthetic gene fragment after restriction digestion by gel extraction.
- Always check the compatible buffer system (► http://nebcloner.neb.com/#!/redigest) and the reaction temperature.

□ **Table 5.12**   DNA ligation with T4 DNA ligase from NEB

Component	Volume [μL]
Water	17 − X − Y
10× T4 DNA ligase buffer	2
Plasmid backbone (pET-19b)	X^a (50 ng)
Insert (Tfu_0901$_{37-466}$)	Y^b
400 U/μL T4 DNA ligase	1
Total	20

[a]Calculated using method below:
Volume of plasmid backbone = 50 ng/(plasmid backbone concentration in ng/μL) = X μL
[b]Calculated using method below:
Length of plasmid backbone = 5717 bp − 17 bp (removed by double digestion) = 5700 bp
Length of insert = 1290 bp
Amount of insert = (50 ng × 3) × (1290 bp/5700 bp) = 34 ng
Volume of insert = 34 ng/(insert concentration in ng/μL) = Y μL

— Whenever possible, use the high-fidelity restriction enzymes (denoted as HF) to reduce star activity (Under non-standard reaction conditions, some restriction enzymes are capable of cleaving sequences which are similar, but not identical, to their defined recognition sequence.).

The molar ratio of insert to plasmid backbone is often kept at 3:1. If you struggle with this calculation, you may use the Ligation Calculator (▶ http://nebiocalculator. neb.com/#!/ligation). Again, we usually do an overnight ligation at 16 °C, and the ligation product (5 μL) can be used directly for bacterial transformation into a cloning strain such as DH5α the next day.

The classical restriction and ligation cloning method remains widely used. But there are some inherent drawbacks with this approach:

— It relies on the availability of two unique restriction sites in the target plasmid for directional cloning.
— Restriction site might leave a 'scar' in the final genetic construct. These undesired extra nucleotide sequences are sometimes translated into additional amino acids.
— The method is time-consuming, requiring multiple PCR and/or gel purification steps.

In the sections below, we will introduce a couple of scarless or seamless cloning methods, which are more advanced.

5.3.2 NEBuilder® HiFi DNA Assembly

NEBuilder® HiFi DNA Assembly is a method for seamless assembly of multiple DNA fragments. It involves the use of different enzymes:

- An exonuclease creates single-stranded 3'-overhangs to facilitate the annealing of fragments that share complementarity at one end (commonly known as the overlap region)
- A polymerase fills in the gaps within each annealed fragment
- A DNA ligase seals the nicks in the assembled DNA

To clone Tfu_0901$_{37-466}$ into pET-19b vector using this approach, we need two PCRs to generate two linear DNA fragments; Tfu_0901$_{37-466}$ and the plasmid backbone pET-19b. This means we require four primers in total, two primers to amplify Tfu_0901$_{37-466}$ from the gDNA (▶ Sects. 5.1.1 and 5.1.2), and two primers to create the linear backbone using pET-19b as the DNA template. Let's design the four primers we need:

- Carry out a virtual cloning and create a plasmid map of pET-19b-Tfu_0901, in which Tfu_0901$_{37-466}$ is 'cloned' into pET-19b between NdeI and BamHI sites.
- Go to the NEBuilder webpage (▶ https://nebuilder.neb.com/#!/)
- Click the '+ NEW FRAGMENT' button
- Under '1. Input source sequence', paste the full sequence of pET-19b-Tfu_0901
- Tick the 'Circular' box
- Under '2. Name/rename fragment', type `Tfu_0901`
- Under '3. Select method for production of linearized fragment', select 'PCR', and enter 'Start base' (`5389`) and 'End base' (`6681`).
- Click the '✓ Add' button
- Two oligonucleotides will be generated for the amplification of Tfu_0901$_{37-466}$ (◘ Table 5.13)
- Click the '+ NEW FRAGMENT' button
- Under '1. Input source sequence', paste the full sequence of pET-19b-Tfu_0901
- Tick the 'Circular' box
- Under '2. Name/rename fragment', type `Backbone`
- Under '3. Select method for production of linearized fragment', select 'PCR', and enter 'Start base' (`6682`) and 'End base' (`5388`).
- Click the '✓ Add' button
- Additional two oligonucleotides will be generated for the amplification of pET-19b (◘ Table 5.13)

NEBuilder® HiFi DNA Assembly relies on the overlapping regions between the two fragments. As such, the primers for Tfu_0901$_{37-466}$ amplification are designed to overlap with those for the pET-19b amplification, as illustrated in ◘ Fig. 5.8.

We have described using the NEBuilder® in a way that should minimize human error, especially for new users. By the time you have tried this tool, you would realise that there are several ways of creating the linear fragments (*i.e.*, PCR, restriction digest and gene synthesis). For instance, it is also possible to use the plasmid backbone fragments prepared by restriction digest. In this case, only two primers will be required for amplifying Tfu_0901$_{37-466}$.

Once you have received your oligos (smallest scale of 0.01 μmol, desalted), you can proceed to creating the two DNA fragments (◘ Tables 5.14 and 5.15) and assembling them (◘ Table 5.16).

The HiFi DNA Assembly mixture is incubated at 50 °C for 15 min, before transforming *E. coli* DH5α using 2 μL of the assembled product.

5

■ **Table 5.13** Oligonucleotides designed by NEBuilder for the amplification of Tfu_0901₃₇₋₄₆₆ and pET-19b vector. Sequences in capital letters represent regions that anneal to the template DNA. Sequences corresponding to Tfu_0901₃₇₋₄₆₆ are coloured in magenta, and those corresponding to pET-19b in blue

Primer	Sequence (5′→3′)	Length [b]	%GC	3′%GC	3′T$_m$	3′T$_a$
Tfu_0901_fwd	caagcatatgGCCGGTCTCAC CGCCACA	28	61	72	74.2	72.0
Tfu_0901_rev	agccggatccTTAGGACTGGA GCTTGCTCCGC	32	63	59	71.8	72.0
Backbone_fwd	ccagtcctaaGGATCCGGCTG CTAACAAAG	30	53	55	66.1	64.3
Backbone_rev	tgagaccggcCATATGCTTGT CGTCGTCG	29	59	53	63.3	64.3

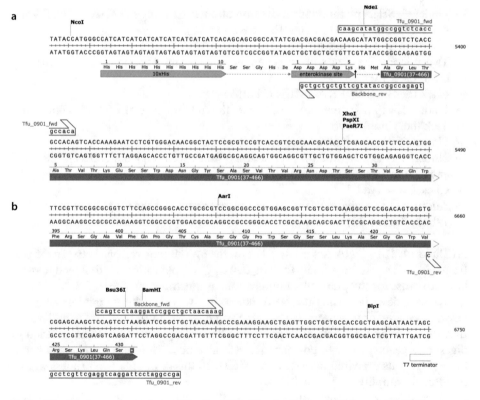

■ **Fig. 5.8** The four primers designed by the NEBuilder for the amplification of Tfu_0901₃₇₋₄₆₆ and plasmid backbone pET-19b. Primer Tfu_0901_fwd overlaps with primer Backbone_rev **a**, and primer Tfu_0901_rev overlaps with primer Backbone_fwd **b**.

◻ **Table 5.14** PCR mixtures for creating linear fragments of Tfu_0901$_{37\text{-}466}$ and pET-19b

Tfu_0901$_{37\text{-}466}$		pET-19b	
Component	Volume [μL]	Component	Volume [μL]
Water	26.5 – X	Water	36.5 – Y
5× Q5 reaction buffer	10	5× Q5 reaction buffer	10
5× Q5 high GC enhancer	10		
10 mM dNTP	1	10 mM dNTP	1
20 μM Tfu_0901_fwd	1	20 μM Backbone_fwd	1
20 μM Tfu_0901_rev	1	20 μM Backbone_rev	1
gDNA	X (100 ng)	pET-19b	Y (100 ng)
2 U/μL Q5 high-fidelity DNA polymerase	0.5	2 U/μL Q5 high-fidelity DNA polymerase	0.5
Total	**50**	**Total**	**50**

◻ **Table 5.15** Thermal cycling conditions for creating linear fragments of Tfu_0901$_{37\text{-}466}$ and pET-19b

Tfu_0901$_{37\text{-}466}$			pET-19b		
Temperature [°C]	Duration	# of cycle	Temperature [°C]	Duration	# of cycle
98	30 sec	1×	98	30 sec	1×
98	8 sec	30×	98	8 sec	30×
72[a]	20 sec		64[c]	20 sec	
72	33 sec[b]		72	2 min 23 sec[d]	
72	2 min	1×	72	2 min	1×
8	∞	–	8	∞	–

[a]Based on the information from ◻ Table 5.13
[b]Calculated using 1290 bp/1000 bp × ((20 sec + 30 sec)/2) = 32.3 sec ≈ 33 sec
[c]Based on the information from ◻ Table 5.13
[d]Calculated using 5712 bp/1000 bp × ((20 sec + 30 sec)/2) = 142.8 sec ≈ 143 sec = 2 min 23 sec

5.3.3 **PTO-QuickStep**

PTO-QuickStep is a seamless cloning method developed in our laboratory. It allows scarless point integration of a DNA sequence (*e.g.*, a gene, a tag) at any position within a target plasmid using only Q5 High-Fidelity DNA Polymerase and DpnI endonuclease. This efficient and cost-effective method consists of two steps: a gene

⬛ **Table 5.16**   Setting up a reaction for two fragments assembly

Component	Volume [µL]
Water	10 – X – Y
pET-19b fragment	X^a (100 ng)
Tfu_0901$_{37-466}$ fragment	Y^b
NEBuilder HiFi DNA Assembly Master Mix	10
Total	20

^{a}Calculated using X = 100 ng/(concentration of pET-19b fragment in ng/µL)
^{b}Calculated using method below:
Length of pET-19b backbone = 5712 bp
Length of Tfu_0901$_{37-466}$ = 1290 bp
Amount of Tfu_0901$_{37-466}$ = (100 ng × 2) × (1290 bp/5712 bp) = 45.2 ng ≈ 46 ng
Volume of Tfu_0901$_{37-466}$ = 46 ng/(Tfu_0901$_{37-466}$ concentration in ng/µL) = Y µL

amplification step using phosphorothioate (PTO) oligos, followed by a megaprimer-based whole-plasmid amplification.

In ⬛ Fig. 5.9 and ⬛ Table 5.17, we demonstrate how the two PTO-oligos required for the amplification of Tfu_0901$_{37-466}$ are designed. The amplified Tfu_0901$_{37-466}$ fragment is then integrated into pET-19b using a second megaprimer PCR:

- Both primers (IntA-Tfu_0901-fwd and IntB-Tfu_0901-rev) are composed of two parts. The regions in magenta bind to the target gene Tfu_0901$_{37-466}$, and regions in blue binds to the target plasmid pET-19b.
- The asterisk signs represent PTO modification at the preceding nucleotide. Each oligo has two PTO modifications.
- These PTO-oligos can be ordered in the smallest synthesis scale (0.01 µmol) and in salt-free format.

Once you have received the PTO-oligos, proceed to gene amplification and megaprimer PCR using reaction mixture and PCR cycling conditions provided in ⬛ Tables 5.18 and 5.19, respectively.

5.4 Clone Verification

We have now shown you three different methods of cloning Tfu_0901$_{37-466}$ into pET-19b. All these methods lead to the same final genetic construct. How do we verify that the cloned sequence is correct? The most direct way is to send the plasmid for DNA sequencing. If we revisit the MCS in ⬛ Fig. 5.2, we can identify potential primers to be used for DNA sequencing. Since the gene Tfu_0901$_{37-466}$ is cloned between NdeI and BamHI sites, we can use the following two standard primers:

- T7 (5′-TAATACGACTCACTATAGGG-3′)
- T7term (5′-CTAGTTATTGCTCAGCGGT-3′)

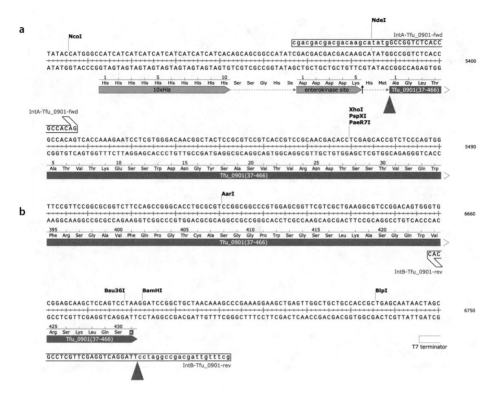

■ **Fig. 5.9** The two PTO-oligos for amplification of Tfu_0901$_{37-466}$. Both oligos IntA-Tfu_0901-fwd **a** and IntB-Tfu_0901-rev **b** are composed of two parts. The 3'-ends of the primers bind to the target gene, and the 5'-ends of the primers carry the target plasmid sequences. The integration sites are indicated with red triangles.

■ **Table 5.17** PTO-oligos for cloning Tfu_0901$_{37-466}$ into pET-19b using PTO-QuickStep. Sequences in magenta are regions that anneal to the target gene Tfu_0901$_{37-466}$, and sequences in blue annealing to pET-19b at the s sites

Primer	Sequence (5'→3')	Length [b]	3'T$_a$ for Tfu_0901[a]	3'T$_a$ for pET-19b[b]
IntA-Tfu_0901-fwd	cgacgacgacg*acaa gcatatg*GCCGGTCT CACCGCCACAG	41	72	69
IntB-Tfu_0901-rev	gctttgttag*cagcc ggatcc*TTAGGACTG GAGCTTGCTCCGCAC	45		

[a]Calculated using NEBTm calculator (▶ https://tmcalculator.neb.com/#!/main) with the sequences coloured in magenta
[b]Calculated using NEBTm calculator (▶ https://tmcalculator.neb.com/#!/main) with the sequences coloured in blue

5

Table 5.18 PCR mixtures for gene amplification and megaprimer PCR

Gene amplification[a]		Megaprimer PCR[b]	
Component	Volume [μL]	Component	Volume [μL]
Water	26.5 – X	Water	38.5 – Y – Z
5× Q5 reaction buffer	10	5× Q5 reaction buffer	10
5× Q5 high GC enhancer	10		
10 mM dNTP	1	10 mM dNTP	1
20 μM IntA-Tfu_0901-fwd	1	PCR product	Y (100 ng)
20 μM IntA-Tfu_0901-fwd	1		
gDNA	X (100 ng)	pET-19b	Z (20 ng)
2 U/μL Q5 high-fidelity DNA polymerase	0.5	2 U/μL Q5 high-fidelity DNA polymerase	0.5
Total	**50**	**Total**	**50**

[a]After PCR, add 6.25 μL of 0.5 M Tris-HCl buffer (pH 9) and 6.25 μL of 100 mM iodine in absolute ethanol directly to the PCR product. After brief mixing by pipetting, incubate the mixture for 5 min at 70 °C then snap-cool on ice. Subsequently, the mixture is purified using a DNA clean-up kit

[b]After PCR, add 40 U of DpnI to the PCR mixture and incubate at 37 °C for 15 min to remove the parental plasmids

Table 5.19 Thermal cycling conditions for gene amplification and megaprimer PCR

Gene amplification			Megaprimer PCR		
Temperature [°C]	Duration	# of cycle	Temperature [°C]	Duration	# of cycle
98	30 sec	1×	98	30 sec	1×
98	8 sec	30×	98	8 sec	25×
72[a]	20 sec		69[c]	20 sec	
72	33 sec[b]		72	2 min 23 sec[d]	
72	2 min	1×	72	2 min	1×
8	∞	–	8	∞	–

[a]Based on the information from Table 5.17

[b]Calculated using 1290 bp/1000 bp × ((20 sec + 30 sec)/2) = 32.3 sec ≈ 33 sec

[c]Based on the information from Table 5.17

[d]Calculated using 5712 bp/1000 bp × ((20 sec + 30 sec)/2) = 142.8 sec ≈ 143 sec = 2 min 23 sec

Most DNA sequencing service providers would have a set of standard primers that includes T7 and T7term.

In situation where your target gene is long (>1 kb), you may need to design additional sequencing primers. In this case, we would recommend the online tools below:

- Eurofins (▶ https://www.eurofinsgenomics.eu/en/ecom/tools/sequencing-primer-design/)
- GenScript (▶ https://www.genscript.com/tools/dna-sequencing-primer-design)

Take-Home Messages

1. NRRL, ATCC, DSMZ, JCM, NCIMB and TBRC are culture collections, from where a microorganism and/or its genomic DNA can be requested or purchased.
2. Cell lysis, protein removal, DNA binding, DNA washing and DNA elution are key steps in a genomic DNA extraction protocol.
3. DNA purity is determined by the A_{260}/A_{280} and A_{260}/A_{230} ratios.
4. The choice of a DNA polymerase is governed by a few factors, which include the application, the polymerase's fidelity and extension rate, among others.
5. Gene synthesis offers numerous advantages (*e.g.*, gene design, codon optimization, and gene mutation *etc*) at a competitive price.
6. There are five categories of molecular cloning approaches: molecular cloning based on complementary overhangs, molecular cloning based on homologous sequences, molecular cloning based on overlapping PCR, molecular cloning based on megaprimers, and molecular cloning based on bridged ligation.
7. Restriction and ligation is a classical molecular cloning method that is still widely used.
8. NEBuilder® HiFi DNA Assembly and PTO-QuickStep are advanced methods for seamless cloning.
9. DNA sequencing is a direct method of verifying a clone after gene cloning.

Exercise

Case Study 4

Cupriavidus necator H16, also commonly known as *Ralstonia eutropha* H16, is a Gram-negative bacterium that naturally accumulates bioplastics [polyhydroxyalkanoate (PHA)] to high percentage of dry cell mass when cultivated under stressed conditions (*e.g.*, nitrogen limitation). The ability of PHA accumulation is endowed by the PHA biosynthesis operon (*phaCAB*) encoding three enzymes (3-ketothiolase, acetoacetyl-CoA reductase and PHA synthase). Your project supervisor would like you to create a PHA-producing *E. coli*, by cloning and expressing the entire *phaCAB* operon from *C. necator* H16.

After your research on various commercial plasmids available, you have decided to clone the *phaCAB* operon into pBAD LIC vector (Addgene catalog #37501).

(a) How would you obtain the *C. necator* H16 strain?
(b) How would you cultivate the *C. necator* H16 strain?
(c) Would you clone the *phaCAB* operon using EcoRV and BamHI sites? Justify your answer.
(d) Design the primers required to clone *phaCAB* operon using NEBuilder® HiFi DNA Assembly kit.
(e) Design the primers required to clone *phaCAB* operon using PTO-QuickStep.

Further Reading

Aschenbrenner J, Marx A (2017) DNA polymerases and biotechnological applications. Curr Opin Biotechnol 48:187–195

Jajesniak P, Tee KL, Wong TS (2019) PTO-QuickStep: a fast and efficient method for cloning random mutagenesis ibraries. Int J Mol Sci 20(16)

Tee KL, Wong TS (2013) Polishing the craft of genetic diversity creation in directed evolution. Biotechnol Adv 31(8):1707–1721

Protein Expression

Contents

© Springer Nature Switzerland AG 2020
T. S. Wong, K. L. Tee, *A Practical Guide to Protein Engineering*, Learning Materials in Biosciences,
https://doi.org/10.1007/978-3-030-56898-6_6

What You Will Learn in This Chapter

In this chapter, we will learn to:

- select and prepare a protein expression medium
- differentiate between a complex medium and an auto-induction medium
- perform protein expression
- optimize protein expression
- understand what an inclusion body is and how to avoid inclusion body formation

The step following gene cloning is to determine expression of our protein of interest. We shall use the genetic construct prepared in ▶ Chap. 5 (*i.e.*, pET-19b-Tfu_0901) to exemplify heterologous protein production in *E. coli*. Typically, we follow the scheme outlined in ◘ Fig. 6.1. Seen in this scheme, there are four key factors affecting the yield of protein expression:

- Expression plasmid with the target gene cloned
- Protein expression host
- Protein expression medium
- Protein expression conditions

We have covered expression plasmid (▶ Sect. 4.2), protein expression host (▶ Sect. 4.1) and how to choose an appropriate host (▶ Sect. 4.3) in ▶ Chap. 4. In the case of pET-19b-Tfu_0901, we will begin with our favourite hosts, *i.e.*, BL21(DE3) and C41(DE3). To reiterate, the DE3 designation means that the respective strains contain the λDE3 lysogen that carries the gene encoding T7 RNA polymerase under the control of a *lac*UV5 promoter (▶ Fig. 4.5). IPTG is therefore required to induce T7 RNA polymerase expression in order to express the recombinant gene cloned downstream of a T7 or a T7*lac* promoter. In pET-19b-Tfu_0901, the expression of Tfu_0901$_{37-466}$ gene is under the control of a T7*lac* promoter.

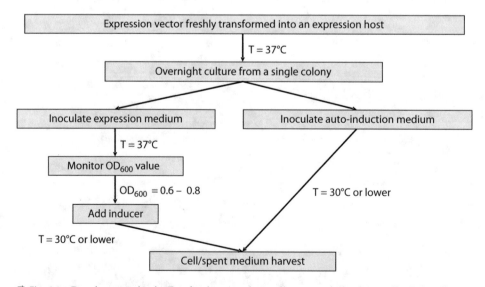

◘ **Fig. 6.1** Protein expression in *E. coli* using complex media or auto-induction media. Depending on the genetic construct, the target protein is present either in the cells or in the spent medium.

In the next two sections, we will discuss protein expression medium (▶ Sect. 6.1) and conditions (▶ Sect. 6.2) in greater detail.

6.1 Protein Expression Medium

For protein engineering, we recommend using the following two types of medium (◻ Table 6.1):
- Complex medium: A complex medium is rich in organic nutrients and minerals. It contains water-soluble extracts of plant/animal tissue (*e.g.*, enzymatically

◻ **Table 6.1** Complex media and their auto-induction counterparts for protein expression

Medium (ordered in increasing 'richness')	Complex medium	Auto-induction medium[a]
Luria-Bertani (LB) broth Miller	10 g/L tryptone 5 g/L yeast extract 10 g/L NaCl	10 g/L tryptone 5 g/L yeast extract 3.3 g/L $(NH_4)_2SO_4$ 6.8 g/L KH_2PO_4 7.1 g/L Na_2HPO_4 0.5 g/L glucose 2 g/L lactose 0.15 g/L $MgSO_4$
2×TY medium	16 g/L tryptone 10 g/L yeast extract 5 g/L NaCl	16 g/L tryptone 10 g/L yeast extract 3.3 g/L $(NH_4)_2SO_4$ 6.8 g/L KH_2PO_4 7.1 g/L Na_2HPO_4 0.5 g/L glucose 2 g/L lactose 0.15 g/L $MgSO_4$
Terrific broth (TB)	12 g/L tryptone 24 g/L yeast extract 4 mL/L glycerol 9.4 g/L KH_2PO_4 2.2 g/L K_2HPO_4	12 g/L tryptone 24 g/L yeast extract 4 mL/L glycerol 3.3 g/L $(NH_4)_2SO_4$ 6.8 g/L KH_2PO_4 7.1 g/L Na_2HPO_4 0.5 g/L glucose 2 g/L lactose 0.15 g/L $MgSO_4$
Super broth	35 g/L tryptone 20 g/L yeast extract 5 g/L NaCl	35 g/L tryptone 20 g/L yeast extract 3.3 g/L $(NH_4)_2SO_4$ 6.8 g/L KH_2PO_4 7.1 g/L Na_2HPO_4 0.5 g/L glucose 2 g/L lactose 0.15 g/L $MgSO_4$

[a]The AIM compositions are obtained/adapted from Formedium AIM products

digested animal proteins such as peptone and tryptone) and yeast extract. The medium is complex because its exact composition is unknown. When a complex medium is used, the inducer IPTG is added to a final concentration of 1 mM when the OD_{600} value reaches between 0.6–0.8.

— Auto-induction medium (AIM): AIM is formulated to cultivate IPTG-inducible expression strains, initially without induction, and then with induction to produce the target protein automatically, usually near saturation at high cell density. A limited concentration of glucose is metabolized preferentially during the initial cell growth phase, which prevents the uptake of lactose until glucose is depleted, usually in mid to late log phase. As the glucose is depleted, lactose is taken up and converted by β-galactosidase into the inducer allolactose. Allolactose causes the release of Lac repressor from its specific binding sites in the DNA (*i.e.*, *lac* operator), thereby inducing the expression of T7 RNA polymerase from the *lac*UV5 promoter and unblocking the T7*lac* promoter to allow expression of the target protein by T7 RNA polymerase. Basically, for every complex medium, there is an AIM counterpart, as shown in ◘ Table 6.1. When cultivating cells in 96-well microplates, we would highly recommend using AIM whenever possible to simplify the process. The key advantages of AIM are:

– Monitoring OD_{600} values to determine point of induction is not required.
– Addition of inducer such as IPTG is not required.
– There is less chance of contamination.
– For the cultivation in 96-well microplates, cells in all wells are induced at about the same OD_{600} value.

Regardless of the type of medium used, an antibiotic is supplemented (*e.g.*, ampicillin in the case of pET-19b-Tfu_0901) to force the cells to keep their plasmids during the course of cultivation. At this point, we would like to introduce the Cold Spring Harbor Protocols (▶ http://cshprotocols.cshlp.org/), which is a lab manual containing many useful recipes and protocols (*e.g.*, how to prepare an antibiotic stock solution, and how to prepare a medium *etc*). For many, if not all, of the media in ◘ Table 6.1, you can purchase as pre-mixed powder or granule from suppliers such as Formedium, Sigma-Aldrich, Thermo Fisher Scientific, and VWR *etc*. Other commercially available complex AIM include Overnight Express™ and MagicMedia™. For those who are interested in more advanced AIM optimization, we recommend reading the excellent work from Studier (PMID 15915565).

Although we seldom use a defined medium in protein engineering, defined medium (*e.g.*, M9) and defined AIM [see Li et al. (PMID 21698378)] for *E. coli* are available.

Depending on the nature of your target protein, additional supplements might be required. For example, 5-aminolevulinic acid (5-ALA, δ-ALA) is commonly added during the expression of heme proteins (*e.g.*, cytochrome P450s). Further, depending on the application of your target protein, you may require animal-free medium (*e.g.*, Vegitone media from Sigma-Aldrich).

6.2 Protein Expression Conditions

More isn't always better! The use of a strong promoter and a high inducer concentration often help to achieve high protein expression but could potentially also lead to protein aggregation. Potential solutions to this problem are:

- Temperature: Lowering the expression temperature to 15–30 °C will improve the solubility of recombinantly expressed proteins. At lower temperatures, cellular processes such as transcription and translation are slowed down, thus leading to reduced protein aggregation. A lower temperature would also reduce protein proteolytic cleavage.
- Inducer concentration: Although 1 mM IPTG is commonly used, a lower concentration of 0.01–0.4 mM can be used to reduce protein aggregation.
- Time: Typically, we would do an overnight protein expression at a lower temperature (*e.g.*, 25 °C or 30 °C for 12–18 hours). You could also use a shorter time at a higher temperature (*e.g.*, 37 °C for 2–4 hours).

6.3 Inclusion Body

The formation of inclusion bodies (IBs) is a common observation during the production of heterologous proteins in bacterial hosts. IBs are exclusively formed by misfolded or unfolded proteins, which aggregate into non-functional protein clusters. In protein engineering, IB formation can present enormous technical challenges, especially when high-throughput protein expression and screening are required. For instance, screening normally entails assessing the activity of a protein, but that would not be possible for non-functional IBs. Tackling protein aggregation is challenging. Strategies to decrease IB formation or IB solubilization often have to be individually tailored for the specific protein target. Here are some of our suggestions:

- Optimize the protein expression conditions (▶ Sect. 6.2)
- Change the expression plasmid (▶ Sect. 4.2) and/or the expression host (▶ Sect. 4.1)
- Solubilization of IB and refolding of protein: This method (◘ Fig. 6.2) is only practical if you are not dealing with a large number of protein variants.
- Identify and mutate the aggregation-prone residues: This is a riskier approach, as a mutation introduced might cause a loss of function. But there are a range of bioinformatic tools to detect aggregation-prone regions or residues, such as:
  - Aggrescan (▶ http://bioinf.uab.es/aggrescan/)
  - PASTA (▶ http://protein.bio.unipd.it/pasta2/)

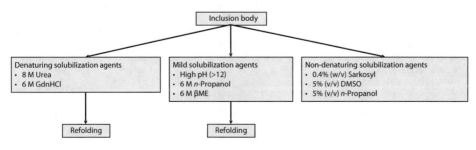

◘ **Fig. 6.2** Different solubilization methods for the recovery of protein from inclusion bodies.

> **Take-Home Messages**
>
> 1. Recombinant protein expression yield is affected by four factors: plasmid, host, medium and expression conditions.
> 2. In protein engineering, complex media and their auto-induction media counterparts are commonly used.
> 3. The use of auto-induction medium avoids the need to monitor cell optical density at 600 nm and to add a chemical inducer.
> 4. Strategies to avoid or reduce protein aggregation include reducing the expression temperature, reducing the inducer concentration and reducing the expression time.
> 5. Inclusion bodies are exclusively formed by misfolded or unfolded proteins, which aggregate into non-functional protein clusters.
> 6. For some proteins, their inclusion bodies can be solubilized and refolded.

Exercise

Case Study 5

Your PhD supervisor has recently come across three new bacterial expression systems:
- Self-inducible expression (SILEX)
- *E. coli* TatExpress strain
- Plasmid system based on triclosan selection

You have been tasked to investigate expressing Tfu_0901$_{37-466}$ using these systems and compare their performance.

(a) What are the advantages of using these new bacterial expression systems?
(b) How would you change your cloning strategy?
(c) How would you change your protein expression protocol?
(d) How would you check for protein expression?

Further Reading

Baneyx F (1999) Recombinant protein expression in *Escherichia coli*. Curr Opin Biotechnol 10(5): 411–421

Kaur J, Kumar A, Kaur J (2018) Strategies for optimization of heterologous protein expression in *E. coli*: roadblocks and reinforcements. Int J Biol Macromol 106:803–822

Studier FW (2005) Protein production by auto-induction in high density shaking cultures. Protein Expr Purif 41(1):207–234

Assay

Contents

© Springer Nature Switzerland AG 2020
T. S. Wong, K. L. Tee, *A Practical Guide to Protein Engineering*, Learning Materials in Biosciences,
https://doi.org/10.1007/978-3-030-56898-6_7

What You Will Learn in This Chapter

In this chapter, we will learn to:

- understand the difference between a direct enzyme assay and an indirect coupled enzyme assay
- understand the difference between a continuous assay and a discontinuous or an end-point assay
- describe various detection modes such as absorbance and fluorescence measurements
- identify chemicals commonly used for absorbance and fluorescence assays
- describe various assay formats and the typical workflow involved
- identify key factors affecting an enzymatic assay
- understand key parameters of an assay
- understand key descriptors of enzyme performance
- choose or develop assays for endoglucanases
- understand the difference between selection and screening

An assay is an analytical method to measure or monitor protein function, which can be enzymatic activity or protein binding. In the context of protein engineering, this can be used to achieve five purposes:

1. To check functional protein expression from the cloned target gene (*e.g.*, the fusion tag is not affecting the enzymatic activity) (▶ Chaps. 4 and 5)
2. To assess and optimize the choice of plasmid, protein expression host, protein expression medium, and protein expression conditions (▶ Chap. 6)
3. To optimize the reaction/binding condition and identify potential inhibitors
4. To evaluate the impact of a mutation on the protein function
5. To identify beneficial variants from a mutant library during directed evolution (▶ Chap. 8)

Assay is the subject of many excellent books. We recommend the Assay Guidance Manual (▶ https://www.ncbi.nlm.nih.gov/books/NBK53196/), which is a free and comprehensive e-book on all aspects of assay. ▶ Chapter 7 will not replicate the information from the Assay Guidance Manual. Instead, we will focus on the key aspects that are pertinent to protein engineering. As with earlier chapters in this book, we will use endo-β-1,4-glucanase Tfu_0901 as a practical example.

7.1 Assay Detection Mode and Formats

For every assay, there needs to be a read-out or a signal that is detectable and quantifiable. This allows us to make comparisons (*e.g.*, wildtype *vs* variants) and to identify beneficial variants from a gene pool. In this section, we will describe various detection mode and formats that can be used for signal detection during assay.

7.1.1 Assay Detection Mode

Rapid development in detection technologies and instrumentation has generated wide-ranging applications. Examples of detection modes are absorbance, fluorescence intensity, fluorescence polarization, time-resolved fluorescence, time-resolved Förster resonance energy transfer, alpha detection, filtered luminescence, flash luminescence, glow luminescence, and nephelometry *etc.* Among these, absorbance assays and fluorescence assays are most commonly used in protein engineering. Both absorbance and fluorescence can be measured using a microplate reader.

7.1.1.1 Absorbance

Absorbance refers to the ability of a substance (called a chromophore) to absorb light of a specific wavelength. Absorbance is calculated using the equation below:

$$A = \log_{10} \frac{I_0}{I}$$

where A = absorbance, I_0 = the intensity of the incident light, and I = the intensity of that light after it passes through the sample. Absorbance is related to transmittance via the relationship below:

$$T = \frac{I}{I_0}$$

$$\%T = 100T$$

$$A = \log_{10} \frac{I_0}{I} = \log_{10} \frac{1}{T} = \log_{10} \frac{100}{\%T} = 2 - \log_{10} \%T$$

where T = transmittance and %T = percent transmittance.
You may recall Lambert-Beer law in ▶ Sect. 2.2.1.1:

$$A = \log_{10} \frac{I_0}{I} = C\varepsilon l$$

where C = concentration, ε = molar extinction coefficient, and l = path length.
    For an enzyme catalysed reaction, activity can be detected using either a direct enzyme assay or an indirect coupled enzyme assay:
— Direct enzyme assay: The substrate or the product of the enzymatic reaction absorbs light at a certain wavelength (▣ Fig. 7.1, Scheme A). Therefore, one can measure a decrease in absorbance (substrate depletion) or an increase in absorbance (product generation). Alternatively, a pseudosubstrate that releases a chromophore can allow direct enzymatic activity measurement (Scheme B). Finally, reaction product can be chemically derivatised to form a coloured compound (Scheme C).
— Indirect coupled enzyme assay: If direct detection of a substrate or a product during the reaction is not feasible, one may consider coupling the enzymatic reaction with another enzymatic reaction that is more easily detectable (Scheme D). In this case, the product of the first enzymatic reaction serves as a substrate for the second enzymatic reaction. This assay type expands the range of enzymatic reactions that can be detected, but at a cost of added complexity to the assay.

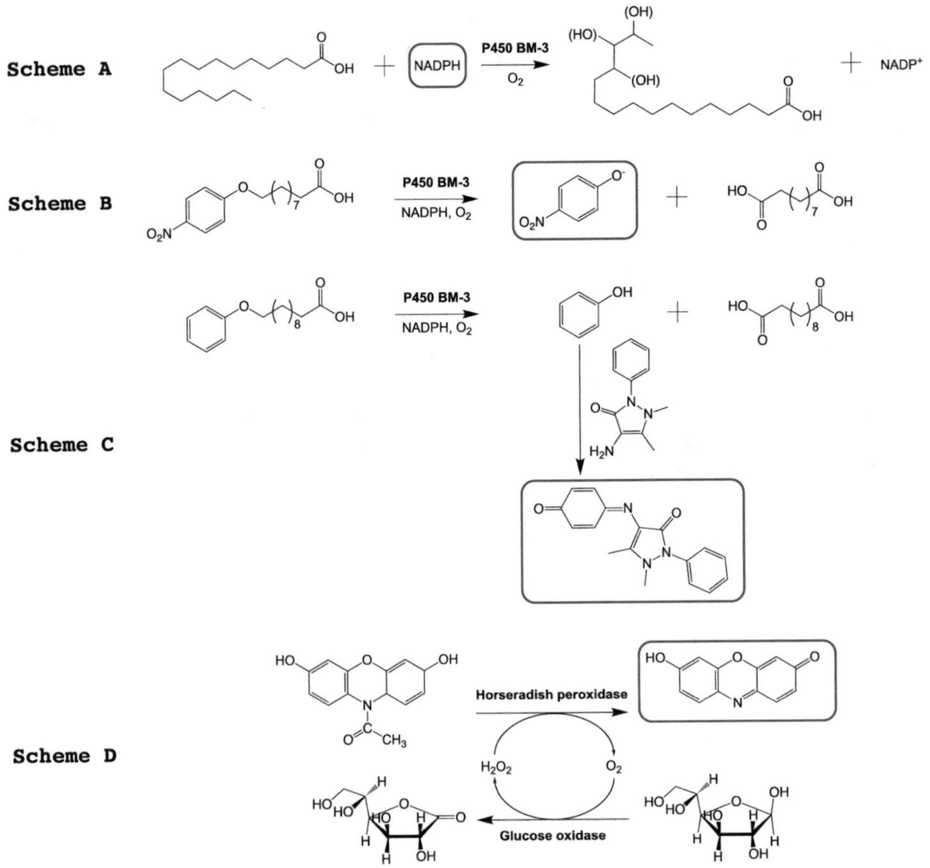

7

◘ Fig. 7.1 Direct enzyme assay *vs* indirect coupled enzyme assay. The chromophores are boxed in blue. In **Scheme A**, a fatty acid is hydroxylated by cytochrome P450 BM-3 at the expense of an NADPH cofactor. The depletion of NADPH can be monitored at 340 nm. In **Scheme B**, a pseudosubstrate 10-(4-nitrophenoxy)capric acid (10-*p*NCA) is used. 10-*p*NCA is a capric acid derivative possessing a *p*-nitrophenol group at its ω-terminus. The *p*-nitrophenolate (*p*NP) anion is released upon hydroxylation by cytochrome P450 BM-3. In **Scheme C**, 11-phenoxyundecanoic acid is hydroxylated by cytochrome P450 BM-3. The terminal phenol released can be derivatised by 4-aminoantipyrine (4-AAP) to yield a red compound 1,5-dimethyl-4-(4-oxo-cyclohexa-2,5-dienylidenamino)-2-phenyl-1,2-dihydro-pyrazol-3-one. In **Scheme D**, oxidation of glucose by glucose oxidase results in the generation of H_2O_2, which is coupled to the conversion of Amplex Red reagent (10-acetyl-3,7-dihydroxyphenoxazine) to resorufin by horseradish peroxidase (HRP). Resorufin is a red-fluorescent product that can be detected spectrophotometrically and fluorometrically.

When we develop an assay in our laboratories, our preference has always been Scheme A > Scheme B > Scheme C > Scheme D, guided by the KISS principle ('Keep it simple and straightforward').

Some chemicals are commonly used in absorbance assays (◘ Table 7.1):

– Cofactor: For enzymes that require a nicotinamide cofactor (*e.g.*, NAD⁺, NADP⁺, NADH or NADPH) as a co-substrate in their reactions, the consumption or the regeneration of the reduced forms can be monitored spectrophotometrically at 340 nm. This is a simple assay for many oxidoreductases such as cytochrome P450s.

◘ **Table 7.1** Commonly used chemicals for absorbance assays in protein engineering

Chemical	Coloured form	Detection wavelength [nm]	Molar extinction coefficient $[M^{-1}cm^{-1}]^a$
10-Acetyl-3,7-dihydroxyphenoxazine (Amplex Red)	Resorufin	571	63000
4-Aminoantipyrine	Quinoneimine	500	12780
4-Amino-3-hydrazino-5-mercapto-1,2,4-triazole (Purpald)	Upon condensation with formaldehyde and oxidation	530–570	7740
2,2'-Azino-bis(3-ethylbenzothiazoline-6-sulfonic acid) (ABTS)	Radical cation	405	36800
Bromocresol green (BCG)	Protonated	610	N/A
Bromothymol blue	Deprotonated	615	N/A
o-Dianisidine	Oxidized	500	7500
2,6-Dichlorobenzoquinone-4-chloroimine (Gibbs reagent)	Indophenol	630	N/A
2,6-Dimethoxyphenol (2,6-DMP)	Coerulignone	469	53200
2,4-Dinitrophenylhydrazine (DNPH)	Dinitrophenyl hydrazone	375	22000
3,5-Dinitrosalicyclic acid (DNS)	3-Amino-5-nitrosalicylic acid (ANSA)	540	N/A
p-Hydroxybenzoic acid hydrazide (PAHBAH)	Osazone complex	410	N/A
Nicotinamide adenine dinucleotide (NAD)	Reduced	340	6220
Nicotinamide adenine dinucleotide phosphate (NADP)	Reduced	340	6300
p-Nitroaniline (pNA)	Free pNA	405	9960
Nitroblue tetrazolium (NBT)-phenazine methosulfate (PMS)	Formazan	585	16000
p-Nitrophenol (pNP)	Deprotonated	405–410	18300–18400
Tetramethylbenzidine (TMB)	Radical cation	653	39000
	Diimine	450	59000
2,3,5-triphenyltetrazolium chloride (TTC)	Formazan	510	N/A

^{a}The molar extinction coefficient is dependent on the wavelength and the local molecular environment

- H$_2$O$_2$ detection agents: For H$_2$O$_2$-producing reactions (*e.g.*, laccases, oxidases, peroxidases), H$_2$O$_2$ can be detected using 2,2'-Azino-bis(3-ethylbenzothiazoline-6-sulfonic acid) (ABTS), 10-acetyl-3,7-dihydroxyphenoxazine (Amplex Red), *o*-dianisidine, tetramethylbenzidine (TMB), 2,6-dimethoxyphenol (2,6-DMP) *etc*, via a coupled reaction typically catalysed by horseradish peroxidase (HRP).
- Nitrobenzene family: Many pseudosubstrates for enzymes such as proteases, cellulases, hydrolases and cytochrome P450s contain a moiety belonging to the nitrobenzene family. When this moiety is cleaved enzymatically, a yellow compound [*e.g.*, *p*-nitrophenol (*p*-NP) and *p*-nitroaniline (*p*-NA)] is released that can be monitored at 405 nm.
- pH indicators: In some enzymatic reactions (*e.g.*, lipases, esterases), fatty acids are released resulting in a change of pH. pH indicators [*e.g.*, bromothymol blue, bromocresol green (BCG)] are used to monitor pH change.
- Redox indicators: For redox reactions, redox indicators such as 2,3,5-triphenyltetrazolium chloride (TTC) and nitroblue tetrazolium (NBT)-phenazine methosulfate (PMS) are used to detect a change in redox potential.
- Derivatizing reagents: Many chemical molecules with functional groups such as –COOH, –OH, –NH and –SH *etc* can be derivatized to form coloured compounds. For examples, 4-amino-3-hydrazino-5-mercapto-1,2,4-triazole (Purpald) and 2,4-dinitrophenylhydrazine (DNPH) react with carbonyl groups (*e.g.*, aldehydes, ketones). 4-Aminoantipyrine and 2,6-dichlorobenzoquinone-4-chloroimine (Gibbs reagent) react with phenolic compounds.
- Chemicals reacting with the reducing end of sugar: This group of chemicals is a subset of derivatizing reagents described above. 3,5-Dinitrosalicyclic acid (DNS) and *p*-hydroxybenzoic acid hydrazide (PAHBAH), for examples, react with sugars (*e.g.*, glucose). Therefore, they are often used to assay polysaccharide-hydrolysing enzymes.

When starting a new protein engineering project, we suggest checking the list of chemical molecules in ▣ Table 7.1 to identify potential candidates that can be used for your enzyme of interest. There are many literatures, protocols and sometimes commercial kits available for these assays. These resources will decrease the time requirement and increase the success rate for your own assay development. Some of these chemicals are suitable for developing a continuous assay, which allows for a continuous, time-dependent monitoring of an enzymatic reaction. However, some chemicals can only be used for a discontinuous assay (also known as an end-point assay), which measures the amount of substrate consumed or product formed after a defined course of an enzymatic reaction.

Let's try to identify suitable assays for Tfu_0901, using the information on the absorbance assay discussed in this section. Endoglucanases, such as Tfu_0901, act by cleaving the internal β-glycosidic bonds in the cellulose chain. There are many activity assays to monitor this reaction, with the partial list provided in ▣ Table 7.2. You may have observed from ▣ Table 7.2, the three schemes adopted for monitoring endoglucanase activity include the use of pseudosubstrates (Scheme B), sugar product derivatization (Scheme C) and indirect coupled enzyme assays (Scheme D). Many of these endoglucanase substrates can be purchased from Sigma-Aldrich, Thermo Fisher Scientific or Megazyme.

◻ **Table 7.2** Absorbance assays for endoglucanases

Substrate	Detection agents	Detection wavelength [nm]	Note
Carboxymethyl cellulose (CMC)	3,5-Dinitrosalicyclic acid (DNS)	540	
Hydroxyethyl cellulose (HEC)	3,5-Dinitrosalicyclic acid (DNS)	540	
AZCL-HEC		590	A pseudosubstrate; HEC dyed with Azurine
AZO-CMC		540	A pseudosubstrate; CMC dyed with Remazolbrilliant Blue
OBR-HEC		550	HEC dyed with Ostazin Brilliant Red H-3B
Barley glucan	3,5-Dinitrosalicyclic acid (DNS)	540	
Avicel	3,5-Dinitrosalicyclic acid (DNS)	540	
4,6-O-(3-Ketobutylidene)-4-nitrophenyl-β-D-cellopentaoside		400	An indirect coupled enzyme assay requiring β-glucosidase
4,6-O-Benzylidene-2-chloro-4-nitrophenyl-β-D-cellotrioside		400	An indirect coupled enzyme assay requiring β-glucosidase
p-Nitrophenyl-D-cellobioside		400	A pseudosubstrate
p-Nitrophenyl-D-glucopyranoside		400	A pseudosubstrate

7.1.1.2 **Fluorescence**

The fluorescence measurement is more sensitive than the absorbance measurement. Fluorescence is the emission of light by a molecule (called fluorophore) that has been excited by a light with a shorter wavelength than the emitted one. Fluorescence intensity detection is the measurement of this emitted light:

$$\Phi = \frac{\# \, emitted \, photons}{\# \, absorbed \, photons}$$

$$I_f = kI_0\Phi(C\varepsilon l)$$

where Φ = fluorescence quantum yield, I_f = fluorescence intensity, and k = proportionality constant attributed to the instrument used for fluorescence measurement.

To work effectively with fluorophores (also called fluorescent probes or fluorescent dyes), you will need to understand their properties (■ Table 7.3). As an example, ■ Fig. 7.2 illustrates the properties of fluorescein, which is used as a pH indicator in some enzymatic assays.

Where do we find the properties of a dye? The properties of a dye are typically provided by its commercial supplier. Further, there are a few excellent databases where you can conduct your search:

— Fluorophores.org (► http://www.fluorophores.tugraz.at/)
— Fluorescence SpectraViewer (► https://www.thermofisher.com/uk/en/home/life-science/cell-analysis/labeling-chemistry/fluorescence-spectraviewer.html#!/)
— Zeiss Filter Assistant (► https://www.micro-shop.zeiss.com/en/uk/shop/filterAssistant/dyes/)

■ Table 7.3 Properties of a fluorophore

Property	Definition	How is it useful for you?
Absorption spectrum	An x-y plot of wavelength *vs* absorbance of a fluorophore.	To determine which excitation wavelength and filter to use in a fluorescence measurement.
Excitation spectrum	An x-y plot of excitation wavelength *vs* number of fluorescence photons generated by a fluorophore. Absorption spectrum is commonly used as a surrogate for the excitation spectrum.	To determine which excitation wavelength and filter to use in a fluorescence measurement.
Emission spectrum	An x-y plot of emission wavelength *vs* number of fluorescence photons generated by a fluorophore.	To determine which emission wavelength and filter to use in a fluorescence measurement.
Molar extinction coefficient (ε)	The capacity for light absorption at a specific wavelength.	A key determinant of the fluorescence output per fluorophore (the 'brightness').
Fluorescence quantum yield (Φ)	Number of fluorescence photons emitted per excitation photon absorbed.	A key determinant of the fluorescence output per fluorophore (the 'brightness').
Photobleaching	Destruction of the excited fluorophore due to photosensitized generation of reactive oxygen species (ROS), particularly singlet oxygen ($^1O_2$).	Results in irreversible loss of fluorescence signal.
Quenching	Loss of fluorescence signal due to short-range interactions between the fluorophore and the local molecular environment, including other fluorophores (self-quenching).	Results in reversible loss of fluorescence signal.

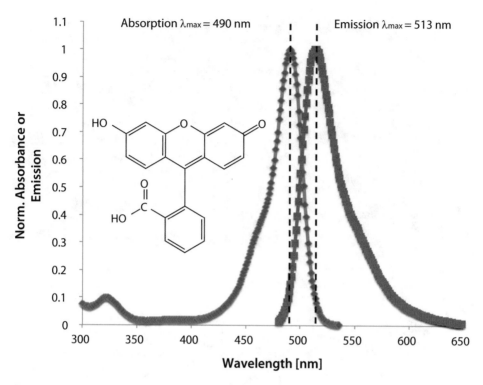

Fig. 7.2 Absorption spectrum and emission spectrum of fluorescein.

In addition to fluorogenic compounds, there is a huge collection of fluorescent proteins [*e.g.*, green fluorescent protein (GFP), red fluorescent protein (RFP)] that you might find useful for your protein engineering project (*e.g.*, as a reporter, as a fusion tag to increase protein solubility or improve protein folding). These proteins, along with their fluorescence properties, can be found in the FPbase (▶ https://www.fpbase.org/).

The concepts we have introduced for absorbance assays (*e.g.*, direct assay *vs* coupled assay, continuous assay *vs* end-point assay) are directly applicable to fluorescence assays. For the same example Tfu_0901, a few fluorescence assays have been reported for endoglucanases (■ Table 7.4 and ■ Fig. 7.3).

7.1.1.3 Others

In addition to the absorbance and the fluorescence assays described above, the hydrolysis products of endoglucanases (*e.g.*, Tfu_0901) can be verified using thin layer chromatography (TLC) and high performance liquid chromatography (HPLC). However, the throughput of TLC and of HPLC is much lower compared to absorbance and fluorescence assays.

□ Table 7.4   Fluorescence assays for endoglucanases

Substrate	Excitation wavelength [nm]	Emission wavelength [nm]	Note
4,6-O-Benzylidene-4-methylumbelliferyl-β-D-cellotrioside (Cellafluor reagent)	365	450	A coupled enzyme assay requiring β-glucosidase
Resorufin cellobioside	571	585	A pseudosubstrate
EnzChek® cellulase substrate	339	452	A pseudosubstrate
5-(4,6-dichlorotriazinyl) aminofluorescein (DTAF)-grafted cellulose	488	515	A pseudosubstrate

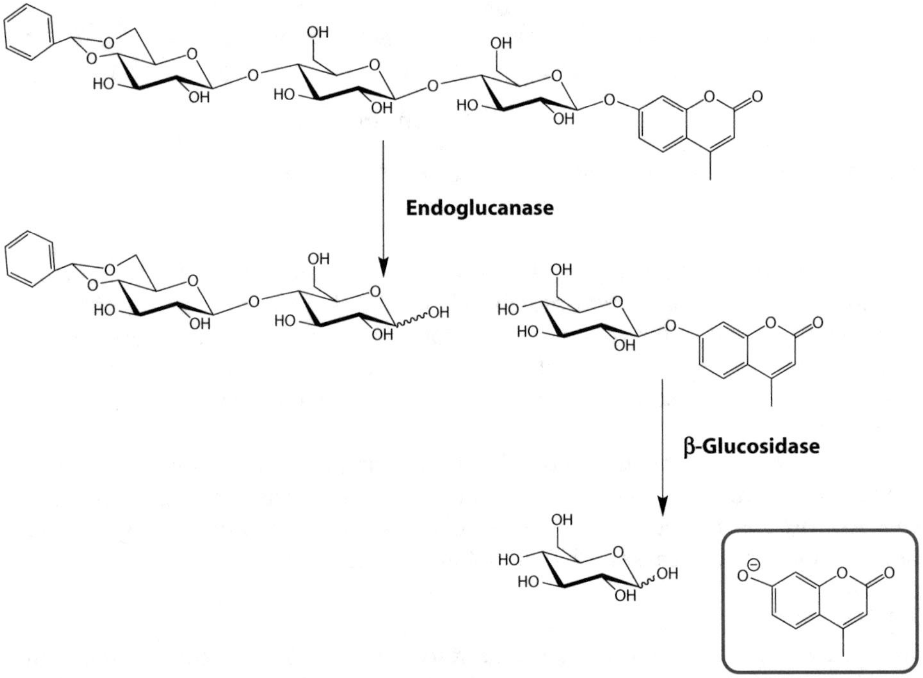

□ Fig. 7.3   Principle of a fluorescence assay based on Cellafluor reagent. Endoglucanase hydrolyses the Cellafluor reagent (4,6-O-benzylidene-4-methylumbelliferyl-β-D-cellotrioside) to 4,6-O-benzylidene-cellobiose and 4-methylumbelliferyl-β-D-glucose. The 4-methylumbelliferyl-β-D-glucose is then cleaved to D-glucose and free 4-methylumbelliferyl (4MU) by the β-glucosidase. The rate of release of 4MU relates directly to the rate of hydrolysis of Cellafluor reagent by endoglucanase.

7.1.2 Assay Formats

In this section, we are going to look at various assay formats (*e.g.*, microplates, agar plates, fluorescence-activated cell sorting *etc*). The choice of assay format is governed by a few factors:

- Enzyme chemistry
- Mode of detection
- Mutant library size
- Research expertise available
- Research facility available

7.1.2.1 Microplate

Microplate assay is the most common format that can be used in research laboratories with a microplate reader. Absorbance assays (▶ Sect. 7.1.1.1) and fluorescence assays (▶ Sect. 7.1.1.2) described above are typically performed in this format.

A broad range of microplates that differ in format, design, material, colour, and surface properties *etc* are available commercially. Identifying the right microplate for your specific application can feel daunting at first. To facilitate this process, ◻ Table 7.5 provides a broad guidance on microplate selection, with protein engineering in mind.

In addition to the standard microplates, there are speciality microplates (*e.g.*, FlowerPlate, baffled 96-deep well plates) that are designed to improve bacterial growth through better mixing and gas-liquid mass transfer.

The typical workflow of using a microplate assay for screening protein variants is depicted in ◻ Fig. 7.4:

- A mutant library (mutant library creation will be covered in ▶ Chap. 8) is created and plated on agar plates.
- Single colonies are picked (using toothpicks, pipette tips, colony pickers, or robotic colony pickers such as QPix, PIXL and Pickolo *etc*) and transferred into microplates containing growth medium supplemented with an appropriate antibiotic.
- When the culture reaches saturation, add glycerol solution, and this master plate is stored in −80°C for further use.
- To begin protein expression, replicate the master plate into a new microplate containing growth medium supplemented with an appropriate antibiotic. This can be done using a pin replicator or pipette tips.
- When the culture reaches saturation, replicate the plate into a new microplate containing medium for protein expression (▶ Sect. 6.1).
- The cell lysate or the spent medium (if the protein is secreted) is used for screening in a fresh microplate (*i.e.*, the assay plate). You can transfer liquid into your assay plates using a multi-channel pipette, a 96-well pipettor, or a liquid handler (high-throughput screening will be covered in ▶ Chap. 9).

7.1.2.2 Agar Plate, Membrane and Digital Imaging

As shown in ◻ Fig. 7.4, a mutant library is originally cultivated on agar plates. Sometimes we can perform a pre-screen directly on the agar plate to differentiate between functional and non-functional variants. This would significantly reduce the

◻ Table 7.5   Guidance on selecting microplates for protein engineering

Features	Options	Application
Format	96-well (8 × 12)	Most commonly used for screening purposes
	384-well (16 × 24)	
	1536-well (32 × 48)	
Materials	Polystyrene	Suitable for screening and spectroscopy applications
	Polypropylene	High chemical resistance and thermal stability. Ideal for microbial cultivation and library storage. Reusable.
	Cycloolefin	Utilised for spectroscopic measurements in the UV range
Pigmentation	Transparent/clear	For colorimetric measurement
	Black	For fluorescence measurement
	White	For fluorescence and luminescence measurement
Surface property	Non-treated	Suitable for biochemical screening
	Non-binding	
Plate bottom	Transparent polystyrene microplate	Colorimetric measurement (>400 nm)
	Polystyrene film bottom microplates with pigmented frame	Spectroscopic measurement (340–400 nm) Fluorescence measurement Luminescence measurement
	UV-transparent cycloolefin film bottom microplates	UV spectroscopy (230–340 nm)
Well design	F-bottom/standard	For precise optical measurement
	U-bottom	For easy and residue-free pipetting
	V-bottom	For precise pipetting and plate centrifugation
Plate height	Standard microplate	Good for microbial cultivation and biochemical screening
	Storage plate/deep well plate	Good for larger volume protein expression

library size in the subsequent microplate assay. In ◻ Table 7.6., we have tabulated some common agar plate assays. These assays are devised based on three principles:
- Clearing zone or halo formation due to substrate depletion.
- Differences in bacterial colony size.
- Differences in bacterial colony colour.

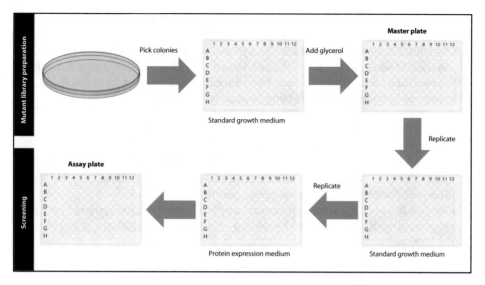

Fig. 7.4 Typical workflow of a microplate assay in protein variant screening.

Table 7.6 Common agar plate assays

Target enzyme	Agar plate preparation	What to observe?
Cellulase	Agar supplemented with CMC and Congo Red	Clearing zone (or halo) indicating cellulolytic activity
Lipase	Agar supplemented with tributyrin or other oil/lipid types	Clearing zone (halo) indicating lipolytic activity
Polyurethanase	Agar supplemented with polyurethane and rhodamine B	Clearing zone (halo) indicating depolymerization activity
Polyesterase	Agar supplemented with Impranii® DLN-SD emulsion, polycaprolactone diol emulsion or polycaprolactone solution	Clearing zone (halo) indicating polyesterase activity
Hydrolase	Standard agar	Smaller colony size indicating cell toxicity due to higher hydrolytic activity
Esterase	Agar supplemented with a pH indicator (*e.g.*, neutral red and crystal violet)	Identification of active esterase variants by the formation of a red colour caused by a pH decrease due to the released acid
Esterase	Minimal medium agar supplemented with glycerol ester	Release of the carbon source glycerol facilitates bacterial growth on minimal media giving rise to larger colonies

In the case of Tfu_0901, we can certainly try the CMC-agar plate assay (■ Fig. 7.5)! The addition of Congo Red helps with visualising the halo formation by providing colour contrast. Congo red can also be replaced by other dyes such as Gram's iodine.

Some agar plate assays are slightly more complex (■ Fig. 7.6), where screening is performed on a membrane and functional variants are identified by visual inspection of colour development or by utilizing advanced digital imaging.

7.1.2.3 Paper-Based Assay

Paper-based assays can be applied in two ways (■ Fig. 7.7):

1. When an enzymatic reaction produces a highly volatile substance (*e.g.*, formaldehyde), one can 'capture' the volatile product by overlaying the microplate with a filter paper wetted with a derivatizing agent (*e.g.*, Purpald).
2. Cell-free lysate or spent medium containing the target protein is spotted on a filter paper, and reaction is initiated by adding substrates (and indicator, if required) to the filter paper.

In both cases, an image of the filter paper can be taken using a camera, a gel imager or even a smart phone, and analysed using a free software such as ImageJ (▶ https://imagej.nih.gov/ij/). ImageJ is compatible with various operating systems (*e.g.*, macOS, Linux and Windows), and open and save various image formats (*e.g.*, TIFF, GIF, JPEG, BMP, and PNG *etc*).

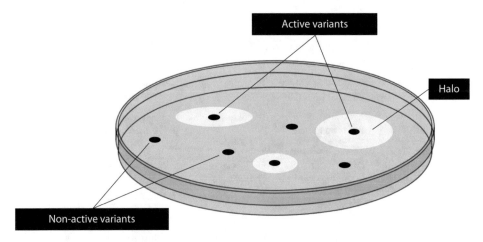

■ **Fig. 7.5**   Typical agar plate assay. Halo formation indicates active variants.

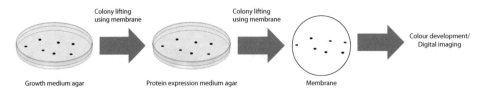

■ **Fig. 7.6**   Assays involving agar plate and membrane. Colonies are lifted using a membrane for transfer from one agar plate to another or for assay.

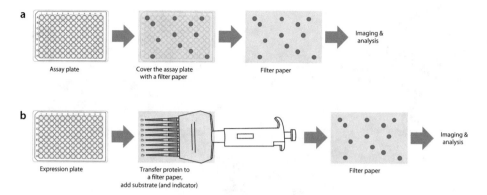

◻ Fig. 7.7 Two types of paper-based assay. **a** The filter paper is used to capture a highly volatile product. **b** The enzymatic reaction is conducted on a filter paper.

While agar plate and paper-based assays are useful as pre-screens to differentiate functional and non-functional clones, they are at best semi-quantitative. Large differences in enzyme activity can be observed by visual inspection of the different halo sizes (agar plate assay) or based on the semi-quantification of spot intensity using imaging software. These two screening formats are not used for precise quantification of enzyme activity.

7.1.2.4 Fluorescence-Activated Cell Sorting (FACS)

Contrary to the three formats above (microplate, agar plate and paper), FACS is technically the most demanding assay format, requiring a specialised equipment (*i.e.*, flow cytometer). Depending on the specific FACS screen, microfluidic systems may also be required to generate monodisperse droplets (or compartments). Generally, there are two types of compartments that can be sorted (◻ Fig. 7.8):

— Whole cells: These are naturally existing compartments.
— Biomimetic compartments: These are man-made compartments, in which an enzymatic reaction takes place. Examples include single emulsion (water-in-oil), double emulsion [water-oil-water (w/o/w) or oil-water-oil (o/w/o)], hydrogel beads, liposomes, and polymersomes. In some applications, cells are entrapped within these biomimetic compartments.

As the name implies, a fluorescence output from the assay is essential for a FACS screen. FACS separates a population of cells or compartments into sub-populations based on their fluorescence signal. This is accomplished in three steps (◻ Fig. 7.8):

— As cells or compartments leave the nozzle tip, they are individually 'interrogated' by a laser.
— Each cell or compartment is given an electrical charge, according to the fluorescence signal within the cell or the compartment.
— Deflection plates attract or repel the charged cells or the compartments into collection tubes.

7

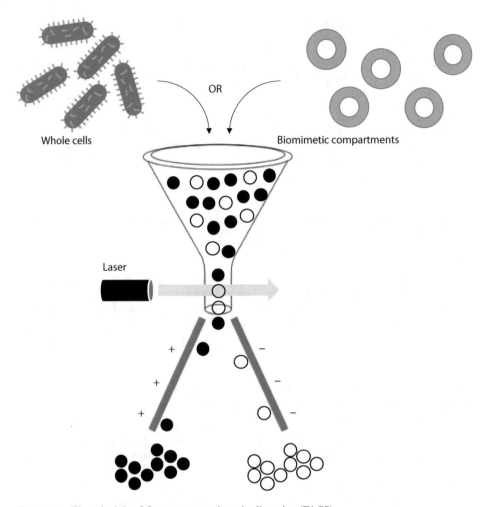

▣ Fig. 7.8 The principle of fluorescence-activated cell sorting (FACS).

An example on how FACS can be applied to screen an enzyme library is shown in ▣ Fig. 7.9. In this example, a pseudosubstrate fluorescein di-β-D-glucopyranoside (FDGlu) is used to assay β-glucosidase activity. To our best knowledge, FACS has not been used to screen endoglucanases. But the assays described in ▣ Table 7.4 can potentially be modified to screen Tfu_0901 by FACS.

7.1.2.5 Pull-Down Assay

Different from the four formats described above, a pull-down assay is mainly used to isolate and identify binding partners of a protein, with the latter used as a 'bait' to 'catch' the binders. A pull-down assay is typically performed as illustrated in ▣ Fig. 7.10. There are four stages in a pull-down assay:

- Protein display
- Binding
- Washing
- Elution

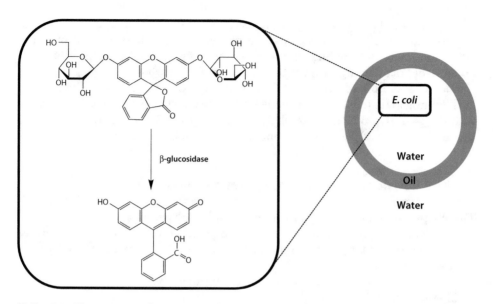

Fig. 7.9 Fluorescence-activated cell sorting of a β-glucosidase library using fluorescein di-β-D-glucopyranoside (FDGlu) as a pseudosubstrate. Cells and the substrates are encapsulated in a w/o/w double emulsion. The nonfluorescent substrate is sequentially hydrolysed by β-glucosidase produced by the cells, first to fluorescein monoglucoside and then to highly fluorescent fluorescein.

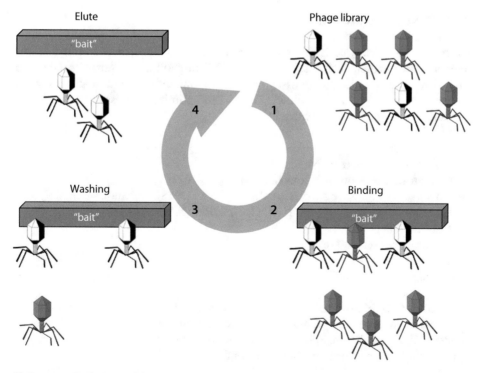

Fig. 7.10 Typical workflow of a pull-down assay for protein displayed on phages.

7

Many display technologies have been reported, since the development of phage display by George P. Smith in 1985. Phage display is utilized in studying protein-ligand interactions, receptor binding sites and in improving or modifying the affinity of proteins for their binding partners. Sir Gregory P. Winter from the Medical Research Council's Laboratory of Molecular Biology and the Centre for Protein Engineering in Cambridge UK applied the technology for the directed evolution of antibodies, with the goal of producing new pharmaceuticals. In ◼ Table 7.7, we have briefly summarized some of the display technologies you can find in the literatures.

7.2 Key Considerations in Assay Development

This chapter have presented you with abundant information on assay. The extensive content is required to appreciate the importance of this topic, the diverse methodologies and the rapid technical advancement. At this point, you may have some questions:
- What are the key considerations in developing an assay?
- How do I optimize my assay?
- Which assay is better?
- What does my assay result tell me?

This next section attempts to answer these questions.

7.2.1 Assay Conditions

Every protein or enzyme is different. Each has its own optimal operating pH and temperature. Further, each displays different binding affinity towards its substrate. Therefore, when developing an assay or applying one for selection or screening, we recommend investigating the following factors to optimize your assay:
- Buffer used and buffering capacity
- pH
- Temperature
- Salt concentration (*e.g.*, NaCl, KCl)
- Concentration of divalent ions (*e.g.*, Mg^{2+}, Ca^{2+}, Zn^{2+})
- Co-solvent to improve substrate solubility (*e.g.*, DMSO, ethanol, methanol)
- Cofactor concentration (*e.g.*, NADH, NADPH, NAD^+, $NADP^+$)
- Enzyme concentration
- Substrate concentration
- Reaction time
- Mixing

We want to re-emphasize some important concepts on buffer and concentration calculations as we have seen students made mistakes in buffer preparation or reagent concentration calculation. While these may seem to be minor errors, they often lead to confusing and even non-reproducible results.

◘ Table 7.7 Display technologies for protein engineering

Category	Display technology	Description
Methods involving natural particles	Phage display	A method for presenting polypeptides on the surface of lysogenic filamentous bacteriophages (*e.g.*, f1, fd, and M13). Proteins and peptides are displayed utilizing the phage proteins pIII or pVIII.
	Baculovirus display	Baculovirus *Autographa californica* multiple nucleopolyhedrovirus (AcMNPV) is used for eukaryotic virus display. Insertion of heterologous peptides or proteins into the viral surface by utilizing the major envelope glycoprotein gp64, or foreign membrane-derived counterparts, allows incorporation of the sequence of interest onto the surface of infected cells and virus particles.
Methods involving cells	Yeast display	*Saccharomyces cerevisiae* offers multiple options for cell surface anchor proteins, including Agα1p, Aga2p, Cwp1p, Cwp2p, Tip1p, Flo1p, Sed1p, YCR89w, and Tir1. Fusing a protein of interest to the C- or N-terminus of an anchor protein typically results in the display of up to 100000 copies of the fusion protein on the cell surface of yeast.
	Bacterial display	Outer membrane proteins (OmpA, OmpC, OmpF, and PhoE), lipoproteins and autotransporters are frequently used as carriers for Gram-negative bacteria surface display.
Cell-free methods	Ribosome display	The key feature of ribosome display is the generation of stable protein-ribosome-mRNA (PRM) complexes via ribosome stalling such that the nascent protein and mRNA remain associated. Two strategies which have been used are: (i) addition of antibiotics such as rifampicin and chloramphenicol (for prokaryotic ribosomes) or cycloheximide (for eukaryotic ribosomes) to halt translation at random, or (ii) deletion of the stop codon, normally recognised by release factors which trigger detachment of the nascent polypeptide, to stall the ribosome at the 3'-end of the mRNA.
	mRNA display	In mRNA display, mRNA molecules bearing a pendant 3' puromycin are translated *in vitro* to generate covalent mRNA-protein fusions.
	Plasmid display	Plasmid display is based on formation of protein-DNA complex, by fusing a peptide or a protein to a DNA-binding domain such as the Lac repressor, NF-κB p50, Oct-1, and GAL4 DNA-binding domain.
	CIS display	CIS display exploits the ability of a DNA replication initiator protein (RepA) to bind exclusively to the template DNA from which it has been expressed, a property called *cis*-activity.
Methods involving man-made vesicles	Liposome display	Liposome display uses cell-sized, giant unilamellar liposomes for cell-free protein synthesis, membrane protein integration and protein function detection for screening.

7.2.1.1 Buffer

Every enzyme has its optimal operating pH. Hence, a buffer solution is used to resist pH changes during an enzymatic reaction when small amounts of acid or base are added. Buffer solutions can be prepared by combining a weak acid and its salt (conjugated base) or, analogously, a weak base and its salt (conjugated acid). For example, phosphate buffer is prepared by mixing NaH_2PO_4 (weak acid) and Na_2HPO_4 (the salt, conjugated base).

We hope you still recall the Henderson-Hasselbalch equation that relates pH to pK_a:

$$pH = pK_a + \log \frac{[conjugate\ base]}{[weak\ acid]}$$

$$pH = pK_a + \log \frac{[A^-]}{[HA]}$$

$$pK_a = -\log K_a$$

$$pH = -\log[H^+]$$

$$\frac{K_a}{[H^+]} = \frac{[A^-]}{[HA]}$$

This equation tells us that $[A^-] = [HA]$, when pH is equal to pK_a. Hence the closer the pH and pK_a are, the higher the buffering capacity. When selecting a buffer solution, choose a buffer solution with a pK_a value close to the target pH level to maximize its buffering capacity.

If you are unsure about buffer preparation, try the Buffer Calculator (▶ https://www.liverpool.ac.uk/pfg/Research/Tools/BuffferCalc/Buffer.html). Whenever you prepare a buffer, do pay attention to the following points:

- When using a pH meter, check that it is calibrated.
- Depending on the buffer, its pH may vary with temperature. One buffer particularly susceptible to changes in temperature is Tris.
- If other components are added to the buffer (e.g., EDTA, DTT, Mg^{2+}, imidazole etc), changes in the pH are expected. Therefore, re-check the pH and adjust if necessary.
- Microbial contamination can occur in a buffer solution. Check for microbial growth before you use your old buffer!
- NaCl can shift the pH of a phosphate buffer.

7.2.1.2 Concentration

Buffer and reagent preparations are inevitable in a wet lab. If you struggle with molarity and dilution, the following two equations may help:

$$M = \frac{Mol\ of\ solute}{L\ of\ solution}$$

$$Dilution\ factor = \frac{Stock\ concentration}{Final\ concentration}$$

As with buffer preparation, there are many excellent online tools to help with these calculations. Good examples include:

— Mass Molarity Calculator: ▶ https://www.sigmaaldrich.com/chemistry/stockroom-reagents/learning-center/technical-library/mass-molarity-calculator.html
— Solution Dilution Calculator: ▶ https://www.sigmaaldrich.com/chemistry/stockroom-reagents/learning-center/technical-library/solution-dilution-calculator.html

7.2.2 Understanding Your Assay

There can be more than one assay available for your specific purpose, as shown for Tfu_0901. When selecting the "right" one to use, it is often practical to consider any limitations in research facilities and choose the simplest assay (*e.g.*, least steps, no heating, no centrifugation). Once you have decided on your assay, you need to understand the operating range of your assay.

Limit of blank (LoB), limit of detection (LoD), and lower limit of quantification (LLoQ) are terms used to describe the smallest analyte concentration that can be reliably measured by an assay (◘ Fig. 7.11). Here are their definitions according to the guidelines published by the Clinical and Laboratory Standards Institute (CLSI; ▶ https://clsi.org/):

— Limit of blank (LoB): The highest apparent analyte concentration expected to be found when replicates of a blank sample containing no analyte are tested.

$$LoB = \mu_{blank} + 1.645(\sigma_{blank})$$

— Limit of detection (LoD): The lowest amount of analyte (measurand) in a sample that can be detected with (stated) probability, although perhaps not quantified as an exact value. LoD is used interchangeably with 'sensitivity', 'analytical sensitivity' and 'detection limit'. The term 'sensitivity' is also loosely used in other ways. For example, 'sensitivity' can also refer to the slope of the calibration curve (◘ Fig. 7.11).

$$LoD = LoB + 1.645(\sigma_{low\ concetration\ sample})$$

— Limit of quantification (LoQ): the lowest amount of analyte (measurand) in a sample that can be quantitatively determined with (stated) acceptable precision and stated, acceptable accuracy, under stated experimental conditions.

$$LoQ \geq LoD$$

In addition to these three limits, there are other assay parameters that you are expected to know (◘ Table 7.8 and ◘ Fig. 7.11).

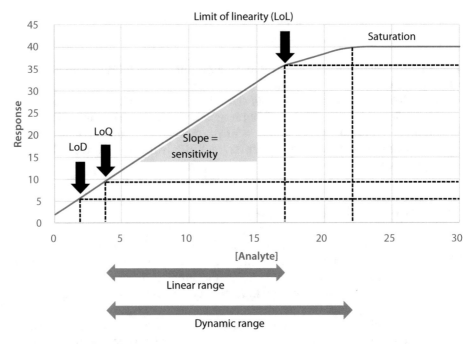

□ **Fig. 7.11**   Assay parameters.

□ **Table 7.8**   Assay parameters

Assay parameters	Key consideration
Specificity	Will the assay detect only the target analyte?
Dynamic range	What are the upper and lower concentrations of the target molecule that the assay can accurately quantify?
Linearity	Is the assay value directly proportional to the concentration of the target molecule?
Interference	Will any component from the cell lysate or assay component interfere with the assay?
Robustness	Can the assay cope with small changes in assay sample, equipment or operator?
Reproducibility	When the assay is repeated, will it give the same outcome? Will another operator be able to replicate the assay?
Accuracy	Does the assay return a value that is in agreement with the expected reference value?
Precision	Is the variability low between replicate measurements?
Throughput	How large is the capacity of the assay? How many clones per day can the assay screen?

7.2.3 **Information You Can Derive from an Assay**

From the perspective of protein engineering, the key objective of conducting an assay is to understand and compare protein variant performances, especially for enzymatic activity. When reading protein engineering literatures, you will come across various ways of describing enzyme performance such as specific activity, turnover number, and maximum velocity *etc*. It is thus useful to understand the meaning of these parameters that describe enzyme performance.

The parameters in ◘ Table 7.9 become easier to comprehend, if you recall the Michaelis-Menten equation (◘ Fig. 7.12):

$$v = \frac{V_{max}C}{K_m + C}$$

where v = enzyme velocity, V_{max} = maximum enzyme velocity, C = substrate concentration and K_m = Michaelis constant.

7.3 **Selection Versus Screening**

Assay can be applied to selection and screening, both methods are used in protein engineering to sift through protein variants, detect functional variants and quantify their activity. These two processes function based on different principles, and hence,

◘ **Table 7.9**　Parameters describing enzyme performance

Parameter	Definition	Unit	Equation
Enzyme unit	Amount of enzyme required to convert 1 µmol of substrate per min	U	$1\ U = 0.0167\ nkat$
Volume activity	Enzyme unit per unit volume	U/mL	
Specific activity	Enzyme unit per unit protein	U/mg	
Enzyme velocity, v	Substrate turnover per unit time	M/s	
Maximum velocity, V_{max}	Substrate turnover per unit time at saturating substrate concentration	M/s	
Turnover number or catalytic constant, k_{cat}	Maximum velocity divided by amount of enzyme	s^{-1}	$k_{cat} = \dfrac{V_{max}}{[E]_0}$
Michaelis constant, K_m	A constant related to the affinity of the enzyme for the substrate or substrate concentration at half-maximal velocity	M	
Catalytic efficiency	Best value to represent the enzyme's overall ability to convert substrate to product	$M^{-1}s^{-1}$	$\dfrac{k_{cat}}{K_m}$

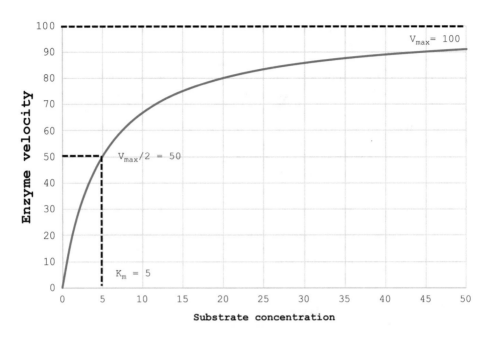

◻ **Fig. 7.12** Relationship between substrate concentration and enzyme velocity according to the Michaelis-Menten equation.

their time requirement, potential throughput and data output vary (◻ Fig. 7.13 and ◻ Table 7.10):

- Selection: Based on the principle of yes/no cut-off, selection provides a growth condition that allows for selective propagation of host cells harbouring those desirable genetic modifications (*e.g.*, mutations that confer growth/survival advantage, auxotrophic complementation), thereby automatically eliminating non-functional variants and enriching those functional ones.
- Screening: Screening, on the other hands, refers to the inspection of every protein variant for the desired property. It involves sampling both functional and non-functional variants.

In practical terms, the throughput offered by selection is unparalleled. It is a quick way to verify functional expression of your protein and identify functional variants from a mutant library during directed evolution (an advanced protein engineering technique discussed in ▶ Chap. 8). An example of a selection system is antibiotic resistance. If functional expression of the target gene confers antibiotic resistance to your clone, the clones can be selected on an agar plate containing the antibiotic. When a selection method is available, our recommendation is to always include it in your protein engineering workflow. The selection method does have one caveat, it does not provide quantitative data for comparison of variants. Hence the screening method can be used to complement selection by providing a more quantitative evaluation of protein function. When a selection method is not available, screening becomes essential during protein engineering for all purposes related to detecting and quantifying protein function. In ▶ Chap. 9, we will discuss high-throughput screening in greater details.

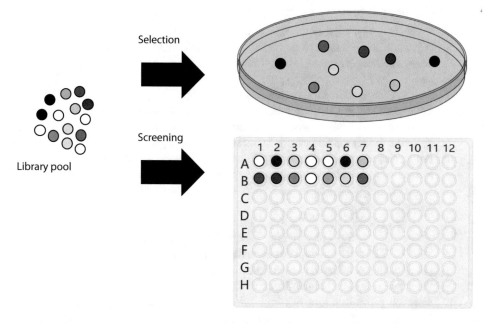

Fig. 7.13 Selection *vs* screening. The colour intensity (from white to black) represents enzymatic activity. Non-functional variants are coloured in white. Functional variants with the highest enzymatic activity are coloured in black, and those with various levels of enzymatic activity in various shades of grey. There are no white clones in selection.

Table 7.10 A comparison between selection and screening

Assay principle	Pre-requisite of the assay	Outcome of the assay	Advantages	Disadvantages
Selection	The function of the target protein must be linked to either growth/survival advantage or auxotrophic complementation	An enriched pool of clones showing improved phenotype	• Unparalleled throughput	• Limited data on individual clone • Often necessitate a secondary phenotypic assay to identify individual clones from the enriched clone pool
Screening	The function of the target protein must be linked to a detectable and quantifiable read-out	Individual clones showing improved phenotype	• Broad compatibility with different assays and detection modes (absorbance, fluorescence, turbidity, HPLC, GC *etc*) • Data output from individual clone	• Time consuming • Throughput is limited by the physical and material constraints required for spatial separation

> **Take-Home Messages**
>
> 1. An assay is an analytical method used to measure or monitor a protein function that can either be an enzymatic activity or a protein-binding ability.
> 2. Every assay needs to provide a read-out or a signal that is detectable and quantifiable.
> 3. An assay forms an integral part of a high-throughput screening workflow to identify protein variants that are superior to their parental clones.
> 4. Absorbance and fluorescence measurements are the two most common assay detection modes used in protein engineering.
> 5. For an enzyme catalysed reaction, enzymatic activity can be detected using either a direct enzyme assay or an indirect coupled enzyme assay.
> 6. A continuous assay allows for a continuous, time-dependent monitoring of an enzymatic reaction.
> 7. A discontinuous assay, or an end-point assay, measures the amount of substrate consumed or product formed after a defined course of an enzymatic reaction.
> 8. Microplate assay is the most common format used in protein engineering.
> 9. Every protein or enzyme has its own optimal operating pH and temperature. Every assay must therefore be optimised for the protein or the enzyme involved.
> 10. Selection provides a growth condition that allows for selective propagation of host cells harbouring those desirable genetic modifications (*e.g.*, mutations that confer growth/survival advantage, auxotrophic complementation), thereby automatically eliminating non-functional variants and enriching those functional ones.
> 11. Screening refers to the inspection of every protein variant for the desired property. It involves sampling both functional and non-functional variants.

Exercise

Case Study 6

Unspecific peroxygenases (UPO) (EC 1.11.2.1) are heme proteins, belonging to the oxidoreductase superfamily. They display both peroxygenase and peroxidase activity. With over 300 identified substrates, UPOs catalyse numerous oxidations including 1- or 2-electron oxygenation and selective oxyfunctionalizations, which make them highly attractive for organic syntheses and industrial biocatalysis.

(a) What is the difference between a peroxygenase and a peroxidase?
(b) Are UPOs dependent on nicotinamide cofactors?
(c) What assay and assay format would you use to monitor the activity of UPO?

Further Reading

Longwell CK, Labanieh L, Cochran JR (2017) High-throughput screening technologies for enzyme engineering. Curr Opin Biotechnol 48:196–202

Silverman L, Campbell R, Broach JR (1998) New assay technologies for high-throughput screening. Curr Opin Chem Biol 2(3):397–403

Sittampalam GS, Kahl SD, Janzen WP (1997) High-throughput screening: advances in assay technologies. Curr Opin Chem Biol 1(3):384–391

Xiao H, Bao Z, Zhao H (2015) High throughput screening and selection methods for directed enzyme evolution. Ind Eng Chem Res 54(16):4011–4020

Gene Mutagenesis

Contents

© Springer Nature Switzerland AG 2020
T. S. Wong, K. L. Tee, *A Practical Guide to Protein Engineering*, Learning Materials in Biosciences,
https://doi.org/10.1007/978-3-030-56898-6_8

What You Will Learn in This Chapter

In this chapter, we will learn to:
- describe the difference between transition and transversion.
- describe the difference between synonymous, non-synonymous and nonsense mutations.
- describe the organizational features of the genetic code and how it affects mutational spectrum and amino acid substitution pattern.
- describe the advantages and disadvantages of directed evolution, rational design and semi-rational design.
- understand the pre-requisites of a successful directed evolution experiment.
- choose a protein engineering strategy.
- choose a gene mutagenesis method.
- create a random mutagenesis library.
- identify essential residues for protein function and target sites for mutagenesis.
- conduct site-directed mutagenesis.
- use OneClick for designing focused mutagenesis experiments.

8

This chapter is all about introducing mutation(s) into the gene of a protein to create a protein variant (some people call it a mutein). We begin by describing the various types of mutations and their consequences on the DNA level as well as on the protein level. This is followed by a brief discussion on the genetic code and three different protein engineering approaches. Finally, we describe different gene mutagenesis techniques, together with a practical demonstration of how mutation(s) can be introduced into the endo-β-1,4-glucanase Tfu_0901$_{37\text{-}466}$.

8.1 Nucleotide Substitution and Amino Acid Substitution

DNA and protein are strings of characters. Nucleotide substitution and amino acid substitution literally means replacing a character by one of its counterparts (3 in the case of a DNA and 19 in the case of a protein). This simplified explanation is a prelude to the content in this section.

8.1.1 Nucleotide Substitution

Each DNA strand is a polynucleotide made up of repeating units called nucleotides. Every nucleotide has three components: a nucleobase (or simply a base), a five-carbon sugar (deoxyribose), and a phosphate group. The nucleobase is either a purine (Pu) [adenine (A) and guanine (G)] or a pyrimidine (Py) [cytosine (C) and thymine (T)]. A nucleobase linked to a sugar is called a nucleoside, while a nucleoside linked to one or more phosphate group(s) is called a nucleotide. The nucleotides are joined to one another by a phosphodiester bond between the sugar moiety of one nucleotide and the phosphate group of the next nucleotide, thereby creating an alternating sugar-phosphate backbone.

The International Union of Pure and Applied Chemistry (IUPAC) has assigned symbols for individual nucleotides or combinations of them (◘ Table 8.1). These are the symbols that you would need when designing or ordering oligos with degen-

Table 8.1 IUPAC symbols for nucleotides and their combinations

Symbol	Description	Bases represented				# of bases represented
A	Adenine	A				1
C	Cytosine		C			1
G	Guanine			G		1
T	Thymine				T	1
U	Uracil				U	1
S	Strong		C	G		2
W	Weak	A			T	2
K	Keto			G	T	2
M	Amino	A	C			2
R	Purine	A		G		2
Y	Pyrimidine		C		T	2
B	Not A (**B** comes after A)		C	G	T	3
D	Not C (**D** comes after C)	A		G	T	3
H	Not G (**H** comes after G)	A	C		T	3
V	Not T and U (**V** comes after T and U)	A	C	G		3
N	Any base	A	C	G	T	4

erate base(s). For example, the pool of oligos 5´-TAATA**C**GACTCACTATAGGG-3´, 5´-TAATA**G**GACTCACTATAGGG-3´ and 5´-TAATA**T**GACTCACTATAGGG-3´ can be represented by the sequence 5´-TAATA**B**GACTCACTATAGGG-3´.

A point mutation is a mutation that affects a single nucleotide of a DNA, often involving the substitution of one nucleotide for another (nucleotide substitution). In total, there are 12 possible nucleotide substitutions ($4 \times 3 = 12$ combinations). These nucleotide substitutions can be divided into:

- Transition (Ts): A nucleotide is changed from a purine to a purine or from a pyrimidine to a pyrimidine.
- Transversion (Tv): A nucleotide is changed from a purine to a pyrimidine or *vice versa*.

According to the rules of nucleotide base pairing, the purine adenine (A) always pairs with the pyrimidine thymine (T), and the pyrimidine cytosine (C) always pairs with the purine guanine (G) in a double-stranded DNA. Hence, an A→G transition in one strand is equivalent to a T→C transition in the other strand. Consequently, the twelve nucleotide substitutions are typically written as six pairs (**□** Table 8.2).

8

■ **Table 8.2** Transitions *vs* transversions

Transitions (Pu↔Pu/Py↔Py)	Transversions (Pu↔Py)
(A→G, T→C) (G→A, C→T)	(A→T, T→A) (A→C, T→G) (G→T, C→A) (G→C, C→G)

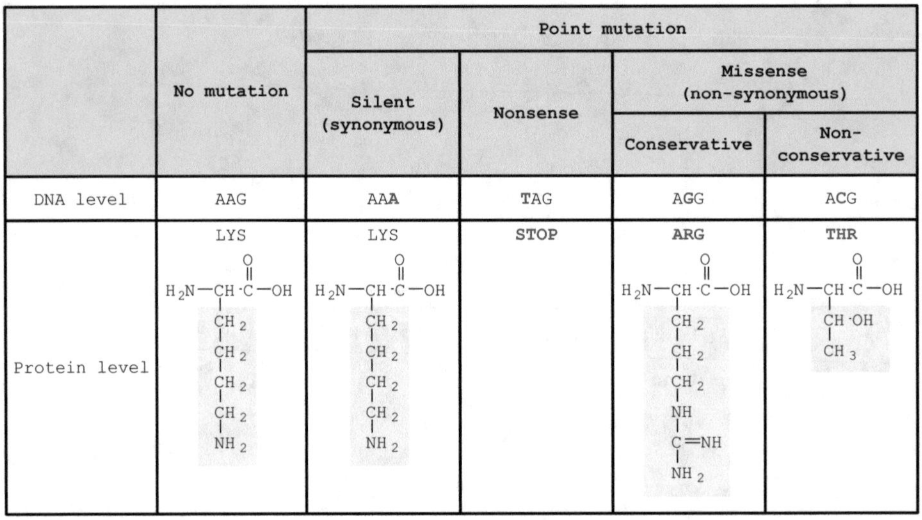

		Point mutation			
	No mutation	Silent (synonymous)	Nonsense	Missense (non-synonymous)	
				Conservative	Non-conservative
DNA level	AAG	AAA	TAG	AGG	ACG
Protein level	LYS $H_2N-CH \cdot C-OH$ $\|$ CH_2 $\|$ CH_2 $\|$ CH_2 $\|$ CH_2 $\|$ NH_2	LYS $H_2N-CH \cdot C-OH$ $\|$ CH_2 $\|$ CH_2 $\|$ CH_2 $\|$ CH_2 $\|$ NH_2	STOP	ARG $H_2N-CH \cdot C-OH$ $\|$ CH_2 $\|$ CH_2 $\|$ CH_2 $\|$ NH $\|$ $C=NH$ $\|$ NH_2	THR $H_2N-CH \cdot C-OH$ $\|$ $CH \cdot OH$ $\|$ CH_3

■ **Fig. 8.1** Point mutations and their effects on amino acid substitutions, using lysine as an example.

8.1.2 Amino Acid Substitution

When a point mutation occurs in a gene, the mutation can be assigned to one of the following categories, according to the consequence of the mutation on the protein (■ Fig. 8.1):

- Synonymous mutation: Also called a silent mutation, the mutation does not change the encoded amino acid.
- Non-synonymous mutation: This mutation changes the amino acid encoded. Non-synonymous mutations are further sub-divided into:
  - Conservative mutation: A mutation that changes an amino acid to another amino acid with similar physicochemical properties (*e.g.*, charge, hydrophobicity and size).
  - Non-conservative mutation: A mutation that changes an amino acid to amino acid with different physicochemical properties.
- Nonsense mutation: A mutation leading to the appearance of a stop codon where previously the codon involved specifies an amino acid. This premature stop codon results in the production of a shortened, and often non-functional, protein product.

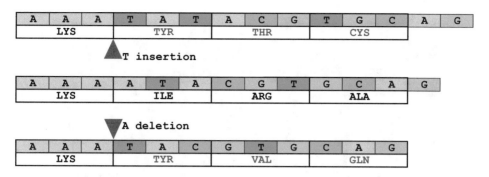

 Fig. 8.2   Frameshift mutations (insertion or deletion).

In addition to point mutation where a single base in the gene is altered, there is also the frameshift mutation. This type of mutation is either an insertion or a deletion involving a number of bases that is not a multiple of three, which consequently disrupts the triplet reading frame of a DNA sequence (Fig. 8.2).

8.2 Organization of the Genetic Code

It is pertinent to discuss the genetic code to properly understand and grasp nucleotide substitution and gene mutagenesis. Many of you would have learned the genetic code in your biochemistry module. But there is more to the code beyond the apparent combination of three nucleotides that form a codon, and this has a huge implication on gene mutagenesis.

The genetic code is a set of rules that defines how the 4-letter code of DNA (A, C, G, T) is translated into the 20-letter code of amino acids (A, C, D, E, F, G, H, I, K, L, M, N, P, Q, R, S, T, V, W, Y), which are the building blocks of proteins. The genetic code is represented by 64 three-letter combinations of nucleotides ($4 \times 4 \times 4 = 4^3 = 64$ permutations) called codons (Fig. 8.3). Each codon corresponds to a specific amino acid or a stop signal. As an example, ATG is a codon that specifies the amino acid methionine (M).

In 1961, Marshall W. Nirenberg (1927–2010) and his postdoctoral fellow Heinrich Matthaei deciphered the first of the 64 triplet codons in the genetic code. In their experiment, an artificial form of RNA consisting entirely of uracil-containing nucleotides, the polyuridylic acid (or poly-U), was added to an *Escherichia coli* cell-free extract to make a protein composed entirely of the amino acid phenylalanine (F). This poly-U experiment showed that UUU specifically coded for phenylalanine (F). Using a series of synthetic RNAs with defined triplets of nucleotides, the entire genetic code was worked out. The 1968 Nobel Prize in Physiology or Medicine was awarded jointly to Marshall W. Nirenberg, Har Gobind Khorana (1922–2011) and Robert W. Holley (1922–1993) 'for their interpretation of the genetic code and its function in protein synthesis'.

The genetic code is nearly universal, with only rare variations reported. Mitochondria, for instance, have an alternative genetic code with slight variations. A single amino acid may be encoded by more than one codon, a phenomenon often

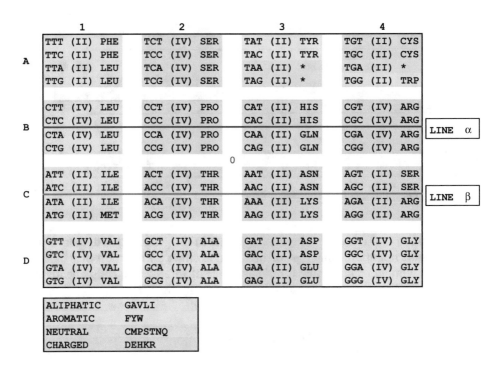

◻ **Fig. 8.3** The genetic code. Amino acids are coloured according to their chemical properties, with aliphatic amino acids in red, aromatic amino acids in green, neutral amino acids in blue and charged amino acids in orange. The subsets of Group IV and Group II are shown as '(IV)' and '(II)', respectively, beside each codon.

described as degeneracy, or redundancy in the genetic code. The genetic code can be sub-divided into two sets of 32 codons (◻ Fig. 8.3):

- Group IV (4-fold degenerate codons): For these codons, a nucleotide substitution at the codon position 3 (cp3) does not alter the encoded amino acid.
- Group II (2-fold degenerate codons): For these codons, all three nucleotides in a codon have to be specified for encoding a specific amino acid or a stop signal.

Contrary to how it may look, the codons are not randomly assigned. If we scrutinize the genetic code, we find the following organizational features:

- Codons encoding the same amino acid vary mainly in their cp3 nucleotide. The variation is often between purines (A/G) or pyrimidines (T/C) in their cp3. As such, a transition mutation at the cp3 nucleotide often does not change the encoded amino acid or likely result in a substitution to an amino acid with a similar chemical property.
- Chemically similar amino acids are encoded by similar codons. For example, codons with a T in their cp2 encode hydrophobic amino acids (F, L, I, M and V), while codons with an A their cp2 encode hydrophilic amino acids (Y, H, Q, N, K, D and E). This explains why the colours in ◻ Fig. 8.3 appear grouped.
- Amino acids within the same biosynthetic pathway are often encoded by codons with an identical cp1 nucleotide (◻ Table 8.3).

◻ **Table 8.3** Biosynthetic pathways of amino acids

Biosynthetic pathway	Amino acids	Identity of 1st nucleotide (number of codons)
Shikimate/aromatic amino acid family	W, Y, F	T (5)
Histidine family	H	C (2)
3-Phosphoglycerate/serine family	S, C, G	T (6), G (4), A (2)
Pyruvate family	A, V, L	G (8), C (4), T (2)
Oxaloacetate/aspartate family	D, N, K, M, T, I	A (14), G (2)
α-Ketoglutarate/glutamate family	E, P, Q, R	C (10), G (2)

— There are one central symmetry (designated as O) and two axial symmetries (denoted as lines α and β) within the genetic code.

— When transversions (A→C, T→G) and/or (G→T, C→A) are applied to a codon at cp1 and cp2, it is possible to change a Group IV codon to a Group II codon or *vice versa* along the central symmetry O. For example, codons in D1 (row D, column 1) can be changed to their symmetrical images, *i.e.*, codons in A4 (row A, column 4), by applying G→T transversion at the cp1 and T→G transversion at the cp2.

— When transversion (A→T, T→A) or (G→C, C→G) is applied to the cp1, it is possible to exchange each Group into itself (Group IV ↔ Group IV or Group II ↔ Group II). Line α exchanges set of codons in row A to its corresponding codons in row C, whereas line β exchanges set of codons in row B to its corresponding codons in row D. For example, codons in A1 (row A, column 1; Group II) can be exchanged to codons in C1 (row C, column 1; Group II) by applying T→A transversion at the cp1. A1 and C1 are symmetrical images when we consider line α as the axis of symmetry.

If we condense the genetic code in ◻ Fig. 8.3 to the 32 codon-format illustrated in ◻ Fig. 8.4, we observe the following features:

— The genetic code is organized in half and half (◻ Fig. 8.4a). The cp3-sensitive (or transversion-sensitive) half encodes more amino acids (15 a.a.), and these amino acids are of diverse physiochemical properties. This is therefore named the pro-diversity half. The other half, however, encodes fewer amino acids (8 a.a.) than the pro-diversity half, whose cp3 nucleotides are also not sensitive to compositional changes.

— The genetic code is partitioned into four quarters (◻ Fig. 8.4b): AT-rich (7 a.a.), GC-rich (4 a.a.), A/Tp1 (6 a.a.) and G/Cp1 (6 a.a.). The AT-rich quarter is the only quarter that encodes both the start codon (M) and the stop codon (*). Therefore, the AT-rich quarter is regarded as the origin of the genetic code evolution.

a

AAR (K)	TAR (*)	GAR (E)	CAR (Q)
AAY (N)	TAY (Y)	GAY (D)	CAY (H)
ATR (M/I)	TTR (L)	GTR (V)	CTR (L)
ATY (I)	TTY (F)	GTY (V)	CTY (L)
AGR (R)	TGR (*/W)	GGR (G)	CGR (R)
AGY (S)	TGY (C)	GGY (G)	CGY (R)
ACR (T)	TCR (S)	GCR (A)	CCR (P)
ACY (T)	TCY (S)	GCY (A)	CCY (P)

Pro-diversity

Pro-robustness

b

AAR (K)	TAR (*)	GAR (E)	CAR (Q)
AAY (N)	TAY (Y)	GAY (D)	CAY (H)
ATR (M/I)	TTR (L)	GTR (V)	CTR (L)
ATY (I)	TTY (F)	GTY (V)	CTY (L)
AGR (R)	TGR (*/W)	GGR (G)	CGR (R)
AGY (S)	TGY (C)	GGY (G)	CGY (R)
ACR (T)	TCR (S)	GCR (A)	CCR (P)
ACY (T)	TCY (S)	GCY (A)	CCY (P)

AT-rich	G/Cp1
A/Tp1	GC-rich

◘ Fig. 8.4 The 32-codon format of the genetic code.

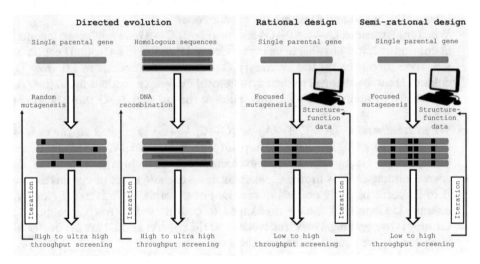

◘ Fig. 8.5 The workflow of directed evolution, rational design and semi-rational design.

8.3 Protein Engineering Approaches

Now that we have discussed nucleotide and amino acid substitution and the genetic code, let's see how mutagenesis is applied in protein engineering. Generally, there are three approaches to protein engineering (◘ Fig. 8.5):

- Directed evolution: This approach mimics the Darwinian evolution. It relies on iterative cycles of genetic diversity creation followed by selection or screening (► Chap. 7), until the desired property is attained. The most common methods to generate genetic diversity include error-prone PCR and DNA recombination.
- Rational design: This is a knowledge-driven process, requiring *a priori* information about the enzyme such as its structure. The knowledge is used to make specific, targeted amino acid mutations. The approach capitalises on the availability of an

increasing number of protein structures, biochemical data, reliable models and computational methods.

— Semi-rational design: This is a combination of rational design and directed evolution. It targets specific residues for saturation mutagenesis or mutagenizing a specific domain/region that is suspected to have a crucial effect on the desired property.

Mutagenesis requirements of the three approaches are different, as highlighted by their comparison in ◘ Table 8.4. The term 'protein sequence space' appears in ◘ Table 8.4. Defined as the maximum number of possible amino acid sequences, it is often used in protein engineering literatures. For a protein of 100 amino acids in length, the protein sequence space is 20^{100}, which is approximately 10^{130}. How large is 10^{130}? The distance between earth and sun is 1.496×10^{11} m, so 10^{130} is astronomically huge!

◘ **Table 8.4**   A comparison of directed evolution, rational design and semi-rational design

	Directed evolution	**Rational design**	**Semi-rational design**
Parental gene	A single gene or a group of homologous sequences	A single gene	A single gene
A priori knowledge requirement	Not required	Required	Required
Genetic diversity creation	Random mutagenesis or DNA recombination	Focused mutagenesis	Focused mutagenesis
Library size	Large	Small	Small to medium
Screening	High to ultra high throughput	Low to high throughput	Low to high throughput
Advantages	• No prior knowledge of the enzyme structure and mechanism is required. • Mutate the entire enzyme, and as such, it is possible to identify mutations distant to the active site that affect the enzymatic activity via allosteric interaction.	• Small library size. • Less time and effort on screening. • Particularly advantageous when there is no high-throughput screening system available.	• Library size is significantly reduced compared to directed evolution. • A larger portion of the protein sequence space is explored compared to rational design.
Disadvantages	• Large library size. • Impossible to explore the full protein sequence space, even with the most powerful selection or screening method. • Time consuming to develop an assay and to screen large library. • Resource intensive.	• *A priori* knowledge is required. • Mutations are mainly targeted at the active site.	• *A priori* knowledge is required. • Mutations are mainly targeted at the active site.

8.3.1 Directed Evolution

Among the three approaches of protein engineering, directed evolution is the most demanding in terms of the large size of its mutant library, and hence, the high capacity needed for screening/selection. It has however been proven time and again to be a very robust process with wide-ranging applications. Let's have a better understanding of directed evolution which will also be relevant for later chapters in this book.

'Survival of the fittest' neatly summarises the principle of directed evolution (◘ Fig. 8.6). This phrase, made famous in the fifth edition of *On the Origin of Species* by British naturalist Charles Darwin published in 1869, suggested that organisms best adjusted to their environment are the most successful in surviving and reproducing. Darwin borrowed the term from English sociologist and philosopher Herbert Spencer, who first used it in his 1864 book *Principles of Biology*.

In 1967, Sol Spiegelman conducted the first extracellular Darwinian experiment with a self-duplicating Qβ RNA molecule (later named the Spiegelman's Monster). The experiment was repeated in 1969 under a defined selection pressure, and Qβ variants resistant to ethidium bromide were isolated. Self-replication offered the possibility of generating optimized molecular phenotypes by 'controlled molecular evolution', according to Manfred Eigen, who was credited the creator of the field of evolutionary biotechnology. In 1984, Manfred Eigen published a theoretical paper outlining the workflow for directed evolution of enzymes by applying the Darwinian logic (◘ Fig. 8.7). He also predicted the possibility to construct an 'evolution reactor' to produce optimized enzymes.

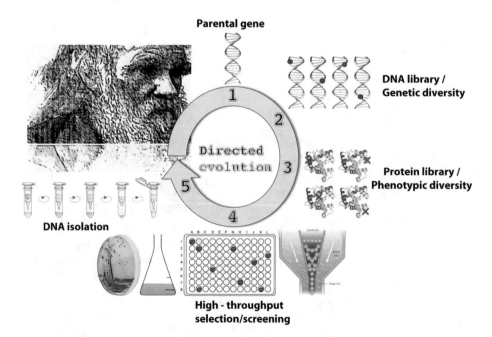

◘ **Fig. 8.6** The workflow of directed evolution.

```
10 PRODUCE A MUTANT SPECTRUM OF SELF-REPRODUCING TEMPLATES
20 SEPARATE AND CLONE INDIVIDUAL MUTANTS
30 AMPLIFY CLONES
40 EXPRESS CLONES
50 TEST FOR OPTIMAL PHENOTYPES
60 IDENTIFY OPTIMAL GENOTYPES
70 RETURN TO 10 WITH A SAMPLE OF OPTIMAL GENOTYPES
```

☐ **Fig. 8.7** The Darwinian logic proposed by Manfred Eigen for molecular evolution.

In 1991, Frances H. Arnold described the experimental implementation of directed evolution of enzymes, following Eigen's proposed workflow. She used random mutagenesis combined with screening to enhance the activity of subtilisin E in the presence of polar organic solvent dimethylformamide (DMF). The method of random mutagenesis by polymerase chain reaction (PCR) that Arnold used, or error-prone PCR (epPCR), was first reported by Lawrence A. Loeb who studied the infidelity of DNA synthesis. Loeb's 1981 paper described the effects of metal ions on the accuracy of DNA polymerases during DNA synthesis, including Mn^{2+}, the most commonly used divalent cation in epPCR. Another prominent contributor to the development and implementation of directed evolution was Willem 'Pim' Stemmer, who introduced a DNA recombination strategy termed 'DNA shuffling' in 1994.

The seminal work of Arnold and Stemmer stimulated the uptake of directed evolution and inspired its application to various enzyme types and biomolecules. In parallel, various methods for creating genetic diversity and for high throughput screening/screening were reported, further fuelling the bloom of protein engineering. In 2018, Frances H. Arnold was awarded the Nobel Prize in Chemistry, along with George P. Smith and Sir Gregory P. Winter.

There are four pre-requisites for a successful directed evolution campaign, formulated by Arnold:

— The phenotype must be biologically feasible, which means there exists a path to get from here to there via an incremental accumulation of beneficial mutations.
— The mutant library must be complex enough to contain the rare beneficial mutations.
— A high-throughput screening/selection system that reflects the phenotype must be available. How quickly the beneficial mutations are identified depends on the speed and the capacity of the screening system.
— The genotype-phenotype linkage must be maintained throughout the entire evolution process. This means any phenotypical improvement can then be traced back to its genetic makeup.

Genotype-phenotype linkage is fundamental to protein engineering. Generally, this linkage can be provided through:

— Spatial separation: Separating the clones in 96-well plates or plating the clones on agar plates.
— Compartmentalization: A cell or a DNA is compartmentalised in man-made vesicles such as emulsion, liposomes and polymersomes.
— Molecular linkage: Linking a protein to its coding RNA or DNA, or displaying a protein on the surface of a cell or a viral particle.

8.4 Mutagenesis

At this point, you should have a good grasp of the concepts of mutations, genetic code and protein engineering approaches. This section looks at the practical aspects of gene mutagenesis. Gene mutagenesis methods are broadly classified into three categories (◻ Fig. 8.8):

- Random mutagenesis: As the name implies, these methods introduce mutations at random positions along a target gene.
- DNA recombination: These methods allow larger gene segments to be exchanged between a group of homologous genes, thereby creating chimeras.
- Focused mutagenesis: These methods insert mutations at specific sites or region within a gene.

When we create a mutant library, there are three types of sequences we want to avoid:

- Parental sequences: We are aiming for protein variants that display superior properties compared to the parent. The presence of parental sequences in a mutant library would only increase the screening burden.
- Sequences encoding non-functional protein variants: More often than not, mutations are deleterious leading to non-functional proteins (*e.g.*, unfolded, prematurely terminated due to a stop codon *etc*). Screening non-functional protein variants is considered unproductive, and therefore, should be minimized and ideally eliminated.
- Repeated sequences: Screening is resource intensive. As such, we would not want to screen the same protein variant twice or thrice. This means repetitive sequences should be avoided, when possible.

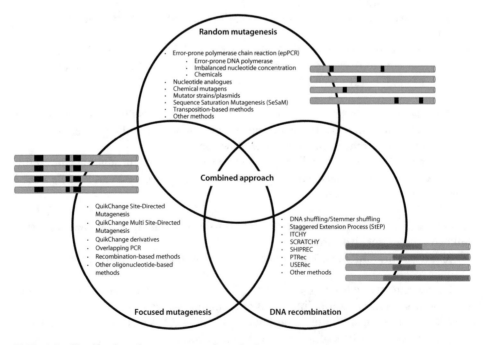

◻ **Fig. 8.8**   Classification of gene mutagenesis methods.

◻ **Table 8.5** An 'ideal' gene mutagenesis method

Criteria	Random mutagenesis	DNA recombination	Focused mutagenesis
Mutation site	Random, no hot spots[a]	Random, no hot spots[a]	Targeted positions only, no off sites
% of parental sequences in the mutant library	0	0	0
% of non-functional sequences in the mutant library	0	0	0
% of repeated sequences in the mutant library	0	0	0
Protein level changes	The original amino acid is changed to one of its 19 counterparts.	All chimeras are in-frame.	The original amino acid is changed to user-defined amino acid(s).

[a]Hot spot, in this context, is defined as the gene position where the probability of being mutated is higher than other positions. The presence of hot spots would mean that mutations are more likely to occur at certain positions, leading to limited genetic diversity

What defines a good gene mutagenesis method? We have summarized what we believe an 'ideal' gene mutagenesis method should look like in ◻ Table 8.5. That said, we need to be realistic in knowing that no method is perfect! The closer the gene mutagenesis method is to the ideal method described in ◻ Table 8.5, the better it is.

In the final two sections of this chapter, we are going to look at how mutant libraries of Tfu_0901$_{37-466}$ can be created. Since we begin with a single gene sequence, we will apply random mutagenesis and focused mutagenesis. DNA recombination must be applied to a group of homologous genes, thus we will leave it out for the time being.

8.4.1 Random Mutagenesis

The simplest random mutagenesis method is arguably the error-prone polymerase chain reaction (epPCR), in which a target gene (*e.g.*, Tfu_0901) is amplified in a non-optimised condition leading to accumulation of errors during DNA replication. Non-optimised PCR condition can be easily created by employing a low fidelity DNA polymerase [*e.g.*, Taq DNA polymerase (▶ Table 5.5) that lacks the 3′→5′ exonuclease activity], along with one or more of the conditions listed below:

– Imbalance nucleotide concentration: In a standard PCR, the concentrations of dATP, dTTP, dGTP and dCTP are kept the same at 200 μM (▶ Table 5.6). Disrupting this nucleotide balance is a common way to 'force' the DNA polymerase to make errors.

— Mg^{2+} and Mn^{2+} concentrations: Reaction buffers for DNA polymerases usually contain 1.5–2.0 mM of Mg^{2+}. Mg^{2+} functions as a cofactor for the activity of DNA polymerases by enabling the incorporation of dNTPs during polymerization. The Mg^{2+} at the enzyme's active site catalyses the phosphodiester bond formation between the 3′-OH of a primer and the phosphate group of a dNTP. Further, Mg^{2+} facilitates the formation of the complex between the primers and DNA templates by stabilizing the negative charges on their phosphate backbones. Higher Mg^{2+} concentration (up to 7 mM) or partial substitution of Mg^{2+} by Mn^{2+} would lead to a higher ratio of incorrect to correct dNTP inserted by the polymerase.

Preparing an epPCR is straightforward, but commercial kits are also available for this purpose (◘ Table 8.6).

8.4.1.1 Mutational Spectrum and Mutation Frequency

When conducting your first epPCR, you will likely have to consider the following:
— Which kit should I use?
— Which epPCR condition should I use?
— How high a mutation frequency should I use?
— How many mutations per gene should I aim for?

To address these questions, we need to first assess the quality of a mutant library created by the different methods/kits. This is reflected in the mutational spectrum of the random mutagenesis methods (◘ Table 8.7) and their mutation frequencies. What is a mutational spectrum? In ► Sect. 8.1.1, you have learned that a nucleotide substitution can either be a transition or a transversion. In total, there are six pairs of nucleotide substitutions. A mutational spectrum gives you the percentage of each pair. Based on these percentages, one can further calculate the mutational bias. In a non-bias method or an ideal method, each pair of nucleotide substitutions would have an equal occurrence of 16.7%. Since there are two pairs of transition and four pairs of transversion, an ideal method would have a Ts to Tv ratio

◘ **Table 8.6** Commercially available epPCR kits

Kit	Supplier	Principle
GeneMorph II random mutagenesis kit	Agilent	Use of low fidelity DNA polymerase (Mutazyme II)
Diversify PCR random mutagenesis kit	TaKaRa	Use of low fidelity DNA polymerase (Taq), imbalanced dNTP and addition of Mn^{2+}
JBS error-prone kit	Jena Bioscience	Use of low fidelity DNA polymerase (Taq), imbalanced dNTP, addition of Mn^{2+}, and higher Mg^{2+} concentration
PickMutant error-prone PCR kit	Canvax	Use of low fidelity DNA polymerase (Taq), imbalanced dNTP and addition of Mn^{2+}

■ **Table 8.7** Mutational spectra of epPCR conditions

Types of mutations	Ideal	GeneMorph II random mutagenesis kit[a]	Taq, 0 mM Mn²⁺, 7 mM Mg²⁺, 0.2 mM dATP, 0.2 mM dGTP, 1 mM dTTP, 1 mM dCTP[b]	Taq, 0.15 mM Mn²⁺, 7 mM Mg²⁺, 0.2 mM dATP, 0.2 mM dGTP, 1 mM dTTP, 1 mM dCTP[b]	Taq, 0.5 mM Mn²⁺, 7 mM Mg²⁺, 0.2 mM dATP, 0.2 mM dGTP, 1 mM dTTP, 1 mM dCTP[b]	Diversify PCR random mutagenesis kit (0 μM Mn²⁺, 40 μM dGTP)[c]	Diversify PCR random mutagenesis kit (640 μM Mn²⁺, 40 μM dGTP)[c]	Diversify PCR random mutagenesis kit (640 μM Mn²⁺, 200 μM dGTP)[c]
Bias indicators								
Ts/Tv	0.5	0.8	1.1	1.2	0.8	0.9	1.3	3.9
AT→GC/ GC→AT	1.0	0.6	1.5	2.1	2.0	4.0	3.7	15.1
A→N, T→N	50%	50.7%	70.0%	80.0%	75.8%	77.8%	77.0%	91.9%
G→N, C→N	50%	43.7%	26.7%	22.5%	19.5%	8.3%	17.8%	7.3%
Transitions, Ts								
A→G, T→C	16.7%	17.5%	30.0%	37.5%	27.6%	33.3%	42.7%	74.0%
G→A, C→T	16.7%	25.5%	20.0%	17.5%	13.6%	8.3%	11.5%	4.9%
Transversions, Tv								
A→T, T→A	16.7%	28.5%	33.3%	37.5%	40.9%	16.7%	26.0%	13.8%
A→C, T→G	16.7%	4.7%	6.7%	5.0%	7.3%	27.8%	8.3%	4.1%
G→C, C→G	16.7%	4.1%	0.0%	0.0%	1.4%	0.0%	0.0%	1.6%

(continued)

Table 8.7 (continued)

Types of mutations	Ideal	GeneMorph II random mutagenesis kit[a]	Taq, 0 mM Mn^{2+}, 7 mM Mg^{2+}, 0.2 mM dATP, 0.2 mM dGTP, 1 mM dTTP, 1 mM dCTP[b]	Taq, 0.15 mM Mn^{2+}, 7 mM Mg^{2+}, 0.2 mM dATP, 0.2 mM dGTP, 1 mM dTTP, 1 mM dCTP[b]	Taq, 0.5 mM Mn^{2+}, 7 mM Mg^{2+}, 0.2 mM dATP, 0.2 mM dGTP, 1 mM dTTP, 1 mM dCTP[b]	Diversify PCR random mutagenesis kit (0 µM Mn^{2+}, 40 µM dGTP)[c]	Diversify PCR random mutagenesis kit (640 µM Mn^{2+}, 40 µM dGTP)[c]	Diversify PCR random mutagenesis kit (640 µM Mn^{2+}, 200 µM dGTP)[c]
G→T, C→A	16.7%	14.1%	6.7%	5.0%	4.5%	0.0%	6.3%	0.8%
Insertions and deletions (InDel)								
Insertions (In)	0.0%	0.7%	0.0%	0.0%	0.3%	2.8%	2.1%	0.0%
Deletion (Del)	0.0%	4.8%	3.3%	0.0%	4.2%	11.1%	3.1%	0.8%
Mutation frequency								
Mutations/kb	N/A	3.0–16.0	1.1	1.9	4.9	2.0	4.6	8.1

[a]Data from the manufacturer's manual
[b]Data from Shafikhani, S. et al., (1997). Generation of large libraries of random mutants in Bacillus subtilis by PCR-based plasmid multimerization. *BioTechniques.* **23**, 304–10
[c]Data from the manufacturer's manual

(Ts/Tv) of 0.5. Based on the data tabulated in ◘ Table 8.7, you would notice the following patterns:

- All the methods are biased towards transitions (Ts/Tv > 0.5).
- Mutations occur more frequently at an 'A' and a 'T' (A→N, T→N > 50.0%)
- Transversion pair (G→C, C→G) has the least occurrence (0.0–5.0%)

If you recall the organization of the genetic code in ▶ Sect. 8.2, a transition-biased method greatly limits the amino acid substitutions we can potentially explore because a transition at cp3 often gives either the same amino acid (synonymous mutation) or an amino acid with a similar chemical property (conservative mutation). This is a key limitation when we rely on the infidelity of a DNA polymerase to introduce DNA replication error. Methods have been developed to bypass this limitation, for example Sequence Saturation Mutagenesis (SeSaM). When you have built sufficient experience as a protein engineer, we strongly encourage you to explore advanced methods such as SeSaM and SeSaM-Tv+.

In addition to the mutational spectrum, it is important to look at the mutation frequency as well. Mutation frequency is expressed in number of mutations per kb of DNA. It is the product of the DNA polymerase error rate (errors per kb of DNA per duplication) and the number of duplications, as shown below:

$$Mutation\ frequency = error\ rate \times d$$

$$Mutation\ frequency\ in\ \% = \frac{Mutation\ frequency}{1000} \times 100\%$$

$$2^d = \frac{PCR\ yield\ in\ ng}{initial\ amount\ of\ template\ DNA\ in\ ng}$$

$$d = \frac{\log_{10}\left(\frac{PCR\ yield\ in\ ng}{initial\ amount\ of\ template\ DNA\ in\ ng}\right)}{\log_{10} 2}$$

$$Mutations\ per\ gene = Gene\ length\ in\ kb \times mutation\ frequency$$

The equations above, along with the mechanisms of introducing DNA replication errors in ▶ Sect. 8.4.1, show that mutation frequency is determined by the following factors:

- Initial amount of template DNA used (affects d)
- Number of PCR cycles (directly determines PCR yield and affects d)
- Concentration of Mn^{2+} (affects error rate)
- Concentration of Mg^{2+} (affects error rate)
- Concentration of each dNTP (affects error rate)
- DNA polymerase used (affects error rate)

In ◘ Fig. 8.9, we show how the mutation frequency varies with the concentrations of Mn^{2+} and dGTP, using the data obtained from the manual of Diversify PCR random mutagenesis kit (TaKaRa).

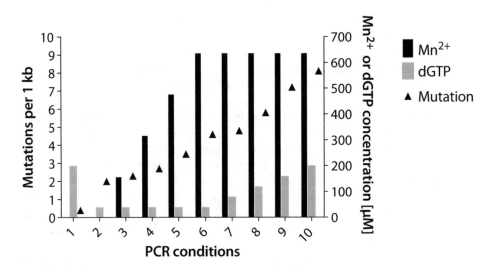

Fig. 8.9 Correlation between mutation frequency (expressed in mutations per 1 kb) and the concentrations of Mn^{2+} and dGTP. Data is obtained from the manual of Diversify PCR random mutagenesis kit (TaKaRa).

When creating a mutant library of a gene (*e.g.*, Tfu_0901_{37-466}), the number of mutations per gene is a better measure of library quality compared to mutation frequency. What is the right number of mutations per gene? Unfortunately, there is no answer to this question. Each gene has its own mutation tolerance (also called the mutational robustness), and the question is best answered experimentally by screening one mutant library. From there, you can then adjust your mutation frequency by changing the epPCR condition. We will revisit this topic in ▶ Chap. 9.

Before we move on, it is important to point out that mutational spectrum and mutation frequency only tell you what happens at the DNA level. They provide no information on amino acid substitution pattern at the protein level. Mutagenesis Assistant Program (MAP) is designed to fill this information gap. MAP translates the nucleotide substitution pattern for different mutagenesis methods into the corresponding amino acid substitution pattern. This online tool is very easy to use, and let's use Tfu_0901_{37-466} as an example:

- Go to the MAP webpage (▶ http://smap.win.biotec.rwth-aachen.de/map3d. html)
- Click 'Submission' in the tool bar
- Paste the coding strand sequence of Tfu_0901_{37-466} in the query box
- Click the 'Submit' button

8.4.1.2 Creating epPCR Libraries of Tfu_0901_{37-466}

We will now demonstrate how to create epPCR libraries of Tfu_0901 using two different kits: GeneMorph II random mutagenesis kit (Agilent) and Diversify PCR random mutagenesis kit (TaKaRa). The gene length of Tfu_0901_{37-466} (▶ Sect. 5.1.3.2) is 1290 bp. To be conservative, we will begin with a low mutation frequency. ◘ Tables 8.8 and 8.9 show the PCR mixture and the thermal cycling conditions that we are going to use, respectively. After the PCR, we can proceed to cloning the gene library using protocols outlined in ▶ Sect. 5.3.1.

◘ Table 8.8 Reaction mixtures for epPCR

GeneMorph II random mutagenesis kit (Agilent)		Diversify PCR random mutagenesis kit (TaKaRa)	
Component	Volume [μL]	Component	Volume [μL]
Water	41.7 – X	Water	40 – Y
10× Mutazyme II reaction buffer	5	10× TITANIUM Taq buffer	5
10 mM dNTP	1	8 mM MnSO$_4$	1
20 μM Fwd primer[a]	0.7 (125 ng)	2 mM dGTP	1
20 μM Rev primer[b]	0.6 (125 ng)	50× Diversify dNTP mix	1
pET-19b-Tfu_0901[c]	X (100 ng)	50× dNTP mix	0
2.5 U/μL Mutazyme II DNA polymerase	1	20 μM Fwd primer[d]	0.5
		20 μM Rev primer[e]	0.5
		pET-19b-Tfu_0901[f]	Y (5.4 ng)[g]
		TITANIUM Taq polymerase	1
Total	**50**	**Total**	**50**

[a]Use the Fwd primer designed in ► Sect. 5.1.3.1. This primer has a molecular weight of 8822.8, calculated using OligoCalc (► http://biotools.nubic.northwestern.edu/OligoCalc.html)
[b]Use the Rev primer designed in ► Sect. 5.1.3.1. This primer has a molecular weight of 10433.8, calculated using OligoCalc (► http://biotools.nubic.northwestern.edu/OligoCalc.html)
[c]Use the construct pET-19b-Tfu_0901 described in ► Sect. 5.3.1
[d]Use the Fwd primer designed in ► Sect. 5.1.3.1
[e]Use the Rev primer designed in ► Sect. 5.1.3.1
[f]Use the construct pET-19b-Tfu_0901 described in ► Sect. 5.3.1
[g]Calculated as follow:
  Gene length of Tfu_0901 = 1290 bp
  Length of pET-19b-Tfu_0901 = 7005 bp
  1 ng of PCR template is equivalent to (7005/1290) × 1 ng of plasmid template = 5.43 ng ≈ 5.4 ng

Using these chosen conditions, we are expecting the mutation frequency and number of mutations per gene shown in ◘ Table 8.10.

8.4.2 Focused Mutagenesis

Unlike random mutagenesis, focused mutagenesis requires *a priori* knowledge to identify target mutation(s) to be introduced into a gene. Since protein structures of Tfu_0901 are available, this is a good example to illustrate how we usually approach focused mutagenesis.

◻ Table 8.9   Thermal cycling conditions for epPCR

GeneMorph II random mutagenesis kit (Agilent)			Diversify PCR random mutagenesis kit (TaKaRa)		
Temperature [°C]	Duration	# of cycle	Temperature [°C]	Duration	# of cycle
95	30 s	1×	94	30 s	1×
95	8 s	23×	94	30 s	25×
60[a]	20 s		68	1 min 18 s[c]	
72	1 min 18 s[b]		68	1 min	1×
72	10 min	1×	8	∞	–
8	∞	–			

[a]Calculated using $T_m - 5\ °C = 65\ °C - 5\ °C = 60\ °C$ (T_m of 65 °C is determined using SnapGene)
[b]Calculated using 1290 bp/1000 bp × 1 min = 1.29 min ≈ 1 min 18 s
[c]Calculated using 1290 bp/1000 bp × 1 min = 1.29 min ≈ 1 min 18 s

◻ Table 8.10   Expected mutation frequency and number of mutations per gene

	GeneMorph II random mutagenesis kit (Agilent)	Diversify PCR random mutagenesis kit (TaKaRa)
Mutation frequency (mutations per kb)	0–4 (on average 2)[a]	2.3[a]
Number of mutations per gene	2 × (1290/1000) = 2.58	2.3 × (1290/1000) = 2.97

[a]Data from manufacturers' manuals

8.4.2.1 **Target Mutation**

In protein engineering, we avoid mutating residues that are critical for protein function (*i.e.*, catalytic residues). The best way to identify catalytic residues is to conduct a quick search in Mechanism and Catalytic Site Atlas (M-CSA; ▶ https://www.ebi.ac.uk/thornton-srv/m-csa/):

- Go to the M-CSA webpage (▶ https://www.ebi.ac.uk/thornton-srv/m-csa/)
- Click 'Search' in the toolbar
- Type `Endoglucanase` in the query box
- Click the 'Search M-CSA' button
- The programme returns three results with M-CSA IDs of 559, 560 and 561.
- If you go through these results, you will realise that catalytic residues of endoglucanase (or cellulase) are negatively charged residues such as glutamic acid (E) and/or aspartic acid (D) (◻ Fig. 8.10).

Step 1

Step 2

☐ **Fig. 8.10**   The mechanism of cellulose cleavage by endoglucanase. In step 1, Glu_A performs a nucleophilic attack on the anomeric carbon. This causes the cleavage of the glycosidic bond assisted by Glu_B protonating the leaving group. In step 2, Glu_B activates a water molecule, which hydrolyses the intermediate.

Typically in focused mutagenesis, we specifically target the following residues, depending on the enzymatic properties we want to improve:
— For enzymatic activity: Active site residues
— For substrate specificity and selectivity: Residues lining the substrate channel or tunnel
— For enzyme stability: Residues that are highly flexible

There are three easy-to-use tools for identifying these residues:
— HotSpot Wizard (▶ https://loschmidt.chemi.muni.cz/hotspotwizard/)
— FireProt (▶ http://loschmidt.chemi.muni.cz/fireprot/)
— CAVER (▶ https://loschmidt.chemi.muni.cz/caverweb/)

Let's try HotSpot Wizard, using Tfu_0901:
— Go to the HotSpot Wizard webpage (▶ http://loschmidt.chemi.muni.cz/hotspot-wizard/)
— Tick the 'Enter PDB code' box
— Enter the PDB ID of Tfu_0901 2CKS in the query box
— Click the 'Process structure' button
— Enter Tfu_0901 as the job title
— Enter your email address
— Tick the licence agreement box
— Click the 'Next' button

- The programme identifies 253E and 355E in Chain A as essential residues (The numbering is based on the 2CKS structure. These essential residues will be eliminated from further calculation).
- Click the 'Submit job' button
- The programme suggests four categories of hot spots to consider in protein engineering:
  - Functional hot spots
  - Correlated hot spots
  - Stability hot spots (structural flexibility)
  - Stability hot spots (sequence consensus)
- If we click on the 'Functional hot spots', the programme identifies five functional hot spots (E192, G298, T360, N390 and D394; numbered based on 2CKS structure) that are all mutable and non-essential for protein function. D394 in Tfu_0901, for example, is located in the catalytic pocket and lysine is the most frequent residue in sequence homologs corresponding to this position. This change from a negatively charged aspartate (D) to a positively charge lysine (K) is significant and potentially interesting for further investigation.

In the last section of this chapter (▶ 8.4.2.3), we will show you how to introduce D394K mutation into Tfu_0901$_{37-466}$.

8.4.2.2 Site-Directed and Multi Site-Directed Mutagenesis

Site-directed mutagenesis was developed by Michael Smith (1932–2000). He was awarded one-half of the 1993 Nobel Prize in Chemistry 'for his fundamental contributions to the establishment of oligonucleotide-based, site-directed mutagenesis and its development for protein studies.' Kary B. Mullis (1944–2019) received the other half of the prize for developing PCR.

As with random mutagenesis, there are commercial kits available for site-directed mutagenesis (◘ Table 8.11). These kits mainly differ in four aspects:
- The DNA polymerase used
- The design of mutagenic oligos
- The number of sites that can be changed simultaneously
- Whether a recombination reaction is required

Of all the methods/kits, QuikChange is most widely used due to its simplicity (◘ Fig. 8.11). QuikChange utilizes a plasmid harbouring the gene of interest and two synthetic oligonucleotide primers, both containing the desired mutation (◘ Fig. 8.11a). The oligonucleotide primers, each complementary to opposite strands of the plasmid, are extended during thermal cycling by a high-fidelity DNA polymerase, without primer displacement. Extension of the oligonucleotide primers generates a mutated daughter plasmid containing staggered nicks. Following whole plasmid amplification, the product is treated with DpnI endonuclease, which is specific for methylated and hemimethylated DNA (target sequence: $5'-Gm^6ATC-3'$), to digest the parental DNA template, leaving the mutation-containing daughter DNA. Important to note, DNA isolated from almost all *E. coli* strains is *dam* methylated, and thus, susceptible to DpnI digestion. The nicked plasmid containing the desired mutations is then transformed into *E. coli* cells.

◘ **Table 8.11** Commercial site-directed mutagenesis kits

Kit	Supplier	DNA polymerase	Mutagenic oligos	Number of sites	A recombination reaction required?
QuikChange II site-directed mutagenesis kit	Agilent	PfuUltra HF DNA polymerase	Overlapping complementary pair	1	✗
QuikChange lightning site-directed mutagenesis kit	Agilent	QuikChange lightning enzyme	Overlapping complementary pair	1	✗
QuikChange XL site-directed mutagenesis kit	Agilent	PfuTurbo DNA polymerase	Overlapping complementary pair	1	✗
QuikChange multi site-directed mutagenesis kit	Agilent	QuikChange multi enzyme blend	1 mutagenic oligo per site	Up to 5	✗
QuikChange lightning multi site-directed mutagenesis kit	Agilent	QuikChange lightning multi enzyme blend	1 mutagenic oligo per site	Up to 5	✗
Q5 site-directed mutagenesis kit	New England Biolabs	Q5 DNA polymerase	Non-overlapping pair	1	✗
Phusion site-directed mutagenesis kit	Thermo Fisher Scientific	Phusion hot start DNA polymerase	Non-overlapping pair	1	✗
GeneArt site-directed mutagenesis system	Thermo Fisher Scientific	AccuPrime Pfx DNA polymerase	Overlapping complementary pair	1	✓
GeneArt site-directed mutagenesis plus system	Thermo Fisher Scientific	AccuPrime Pfx DNA polymerase	1 overlapping complementary pair per site	Up to 3	✓

Within the QuikChange product family, there are kits for multi site-directed mutagenesis. Their principles are very similar (◘ Fig. 8.11b), but all oligonucleotides are designed to bind the same strand of the template DNA. A high-fidelity DNA polymerase is used to extend the mutagenic primers without primer displacement, generating ds-DNA molecules with one strand bearing multiple mutations and containing nicks. The nicks are sealed by a ligase. Following DpnI digestion, the reaction mixture that is enriched with mutated single-stranded DNA, is transformed into *E. coli* cells where the mutant closed circle ssDNA is converted into duplex form *in vivo*.

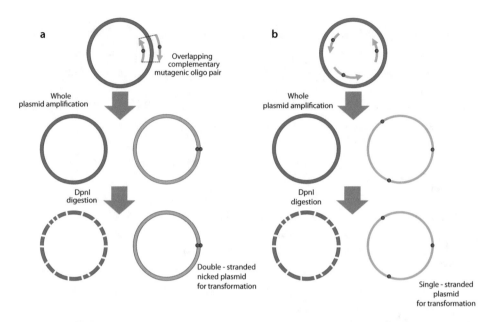

a

Overlapping
complementary
mutagenic oligo pair

Whole
plasmid amplification

DpnI
digestion

Double - stranded
nicked plasmid
for transformation

b

Whole
plasmid amplification

DpnI
digestion

Single - stranded
plasmid
for transformation

□ **Fig. 8.11**   QuikChange site-directed mutagenesis **a** and QuikChange multi site-directed mutagenesis **b**.

Whole plasmid amplification followed by DpnI digest for parental template removal is a widely adopted strategy for site-directed mutagenesis. It also inspires the PTO-QuickStep cloning method we have learned in ▶ Sect. 5.3.3. But the design of mutagenic oligos in QuikChange is not optimal for two reasons:

- The two oligonucleotides in the primer pair are complementary to one another. As such, they have the tendency to form a primer dimer instead of a primer-template complex necessary for primer extension by DNA polymerase. This directly translates to lower PCR efficiency.
- The two oligonucleotides overlap in the template sequence they anneal to. Therefore, the whole plasmid amplification, as depicted in □ Fig. 8.11, is a linear amplification, as opposed to an exponential amplification that is far more efficient.

There are two ways to overcome these two drawbacks:
- Use a partially overlapping and complementary oligo pair, as shown in □ Fig. 8.12.
- Conduct a two-stage PCR:
  - Stage 1: Do two parallel PCR reactions, with only one mutagenic oligo in each reaction. Stage 1 is a short PCR run of 10 cycles.
  - Stage 2: Combine the two reactions from Stage 1, and run an additional 20 cycles.

8.4.2.3 Site-Directed Mutagenesis of Tfu_0901$_{37-466}$

Designing and setting up a reaction for site-directed mutagenesis is very simple. We have developed a publicly accessible online tool called OneClick for this purpose. The programme has incorporated information of most DNA polymerases and plasmid systems. It gives you the flexibility in choosing the type of mutagenic primers you can

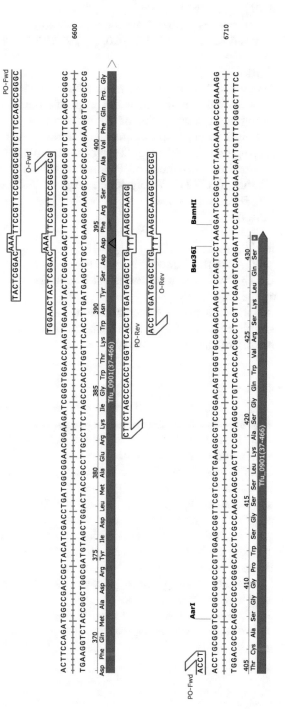

Fig. 8.12 Overlapping primers (O-Fwd and O-Rev) and partially overlapping primers (PO-Fwd and PO-Rev), designed by OneClick, for introducing D394K mutation into Tfu_0901$_{37-466}$. Mutation site (residue 394) is indicated by a red triangle.

design (*e.g.*, overlapping, partially overlapping, non-overlapping), as well as the PCR modes (one-stage or two-stage).

Let's us show you how OneClick can simplify your focused mutagenesis experiment by using Tfu_0901$_{37-466}$ as an example:

- Go to the OneClick webpage (▶ http://www.oneclick-mutagenesis.com/)
- Paste the coding strand sequence of Tfu_0901$_{37-466}$ GCCGGT......TCCTAA in the query box
- Click the 'Begin' button
- Type 394 in the 'Amino acid position to be mutated'
- Tick 'Substitution to a specific amino acid' under 'Type of amino acid substitution'
- Select 'K – Lysine' in the dropdown menu of 'Target amino acid'
- Select 'Escherichia coli B' in the dropdown menu of 'Protein expression host'
- Tick 'Partially overlapping primers' under 'Type of mutagenic primers'
- Select 'pET-19b' in the dropdown menu of 'Plasmid in which the gene is cloned'
- Select 'PfuUltra II Fusion HS DNA Polymerase (Agilent Technologies)' in the dropdown menu of 'DNA polymerase'
- Click the 'Next' button
- Tick 'Two-stage PCR'
- Click the 'OneClick' button
- The programme will return the following information:
  - The mutagenic oligos you need to order (see ◘ Fig. 8.12)
  - How to prepare the PCR mixture
  - How to programme the thermal cycling conditions
  - The type of agar plate you need to use during bacterial transformation
  - The genetic diversity of the mutant library

Take-Home Messages

1. A transition (Ts) is a nucleotide substitution from a purine to a purine or from a pyrimidine to a pyrimidine. There are two pairs of transitions.
2. A transversion (Tv) is a nucleotide substitution from a purine to a pyrimidine or *vice versa*. There are four pairs of transversions.
3. The mutation that does not change the encoded amino acid is known as a synonymous mutation or a silent mutation.
4. Non-synonymous mutation changes the amino acid encoded. A mutation that changes an amino acid to another amino acid with similar physicochemical properties (*e.g.*, charge, hydrophobicity and size) is a conservative mutation. On the other hand, a mutation that changes an amino acid to amino acid with different physicochemical properties is a non-conservative mutation.
5. Nonsense mutation leads to the appearance of a stop codon where previously the codon involved specifies an amino acid.
6. The genetic code is degenerate or redundant. Its organization is not random.
7. There are three protein engineering approaches, directed evolution, rational design and semi-rational design.

8. Directed evolution is based on the Darwinian evolution process, involving iterative cycles of mutagenesis and screening/selection. It does not require *a priori* knowledge of the enzyme structure and mechanism.

9. Frances H. Arnold outlined four key requirements for a successful directed evolution experiment, which cover the protein property that can be evolved, the quality of a mutant library, the capacity of a screening/selection system and the genotype-phenotype linkage.

10. Gene mutagenesis methods can be classified into three categories: random mutagenesis, focused mutagenesis and DNA recombination.

11. Error-prone polymerase chain reaction (epPCR) is most widely used for random mutagenesis.

12. QuikChange method is most widely used for site-directed and multi site-directed mutagenesis.

13. OneClick is a tool developed specifically for designing focused mutagenesis experiments.

Exercise

Case Study 7

Cutinases from *Thermobifida fusca* YX (Tfu_0883; UniProt Q47RJ6) and *Thermobifida fusca* KW3 (TfCut2; UniProt E5BBQ3) are known to degrade polyethylene terephthalate (PET). Site-directed mutagenesis experiments showed that the double mutation Q132A/T101A and single mutation G62A increased the PET-hydrolysing activities of Tfu_0883 and TfCut2, respectively. Your supervisor has asked you to investigate these protein variants.

(a) Describe your cloning and protein expression strategies for Tfu_0883 and TfCut2.
(b) How would you introduce double mutation Q132A/T101A into Tfu_0883?
(c) How would you introduce single mutation G62A into TfCut2?
(d) How would you introduce random mutations into Tfu_0883?

Further Reading

Arnold FH (2019) Innovation by evolution: bringing new chemistry to life (Nobel Lecture). Angew Chem Int Ed Engl 58(41):14420–14426

Knight RD, Freeland SJ, Landweber LF (1999) Selection, history and chemistry: the three faces of the genetic code. Trends Biochem Sci 24(6):241–247

Liu CC, Schultz PG (2010) Adding new chemistries to the genetic code. Annu Rev. Biochem 79:413–444

Sirover MA, Loeb LA (1976) Metal-induced infidelity during DNA synthesis. Proc Natl Acad Sci U S A 73(7):2331–2335

Tee KL, Wong TS (2013) Polishing the craft of genetic diversity creation in directed evolution. Biotechnol Adv 31(8):1707–1721

Verma R, Wong TS, Schwaneberg U, Roccatano D (2014) The mutagenesis assistant program. Methods Mol Biol 1179:279–290

Wang W, Malcolm BA (1999) Two-stage PCR protocol allowing introduction of multiple mutations, deletions and insertions using QuikChange site-directed mutagenesis. BioTechniques 26(4):680–682

Wong TS, Tee KL, Hauer B, Schwaneberg U (2004) Sequence saturation mutagenesis (SeSaM): a novel method for directed evolution. Nucleic Acids Res 32(3):e26

Wong TS, Zhurina D, Schwaneberg U (2006) The diversity challenge in directed protein evolution. Comb Chem High Throughput Screen 9(4):271–288

8

High-Throughput Screening (HTS)

Contents

© Springer Nature Switzerland AG 2020
T. S. Wong, K. L. Tee, *A Practical Guide to Protein Engineering*, Learning Materials in Biosciences,
https://doi.org/10.1007/978-3-030-56898-6_9

What You Will Learn in This Chapter

In this chapter, we will learn to:

— identify the main sources of errors in a high-throughput screening and propose solutions to these problems.
— determine if a HTS system is suitable for screening a mutant library.
— analyse and present the data from a HTS.
— calculate mean, standard deviation and coefficient of variation from a dataset.
— understand the difference between an apparent coefficient of variation and a true coefficient of variation.
— adjust HTS system as the protein variants improve.

In ▶ Chap. 7, you have learned a great deal about assay. Students often equate 'an assay' to 'screening', which is incorrect. An assay is an analytical test to measure a biological function (*e.g.*, an enzymatic activity). Screening, on the other hand, is a sifting process that involves the use of an assay. As screening occurs in a high-throughput manner, you are more likely to face technical challenges due to parallelization and miniaturization of the assay.

We introduced different assay formats in ▶ Chap. 7, but we will focus on HTS of enzymatic activity in 96-well format in this chapter, which is arguably the most versatile and widely used format in protein engineering. Recall the typical workflow of a microplate assay in protein variant screening (▶ Fig. 7.4). It will be useful in this chapter, as we discuss four important questions in HTS:

— What are the common technical challenges and solutions in HTS?
— How do I design the right assay conditions during screening?
— How do I determine the reliability of my HTS process?
— How do I analyse and present my HTS data for the mutant library?

9.1 Practical Aspects of HTS

The HTS process typically requires three key activities, cell cultivation, cell lysis and enzymatic assay. While these activities are commonly performed in many laboratories, HTS requires performing thousands or more of these experiments in parallel and in miniaturized (*i.e.*, microliter) format. The parallelization and scaling down of these experiments are not trivial tasks. Errors will multiply with the high sample number, inaccuracy will magnify with scaled down volumes and consistency will become harder to achieve. Despite these challenges, we know that a reliable HTS process can be achieved, as we and many others have demonstrated. Let's look at some of these technical challenges and their solutions as well as useful practices during HTS.

9.1.1 Liquid Handling

The entire HTS process involves frequent pipetting of small liquid volume. If we use Corning 96-well microplates as examples, total well volumes are 360 μL, 330 μL and 320 μL for F bottom plates, U bottom plates and V bottom plates, respectively. The recommended working volume of each well is 100–200 μL.

☐ Table 9.1 Liquid handling tools compatible with 96-well microplates

Tool	Example of supplier	Price	Number of channels	Typical volume range
Multichannel pipette	Eppendorf	+	8 or 12	0.5–10 µL 10–100 µL 30–300 µL 120–1200 µL
Electronic multichannel pipette	Eppendorf	+	8 or 12	0.5–10 µL 5–100 µL 15–300 µL 50–1200 µL
Microplate dispenser	Integra	++	8	0.5–9999 µL
Liquidator	Mettler Toledo	++	96	0.5–20 µL 5–200 µL
Semi-automated pipetting system	Mettler Toledo	++	96	0.5–20 µL 5–200 µL
Automated liquid handling system	Beckman Coulter	+++	96	1–300 µL 5–1200 µL

A range of tools have been developed by various suppliers for high-throughput liquid handling, and they are compatible with the 8×12 format of 96-well microplates (☐ Table 9.1).

Accuracy and precision of liquid handling are both important in HTS. Accuracy refers to the closeness of a measurement to the actual value, while precision is the closeness among repetitive measurements. To avoid or reduce pipetting errors, we would make the following recommendations:

— Avoid pipetting small volume (*e.g.*, <10 µL) by diluting the reagents to be transferred.
— Reduce the number of pipetting steps by combining the components (*e.g.*, premix the culture medium with antibiotic).

9.1.2 Internal Control

The key function of screening is to sift through a library of protein variants to identify those that are superior to their parent in enzyme performance. During HTS, the relative performance is the primary information that we seek. Therefore, the parental clone must be present in every microplate to serve as the benchmark for comparison (*i.e.*, internal control). In fact, wells B2, E6 and G11 are typically inoculated with parental clones (☐ Fig. 9.1). Three wells are selected to not only serve as benchmark but also as a practical means to evaluate the precision of HTS process (*i.e.*, how close the measurements are among these repeats; ► Sect. 9.4). The selected wells are scattered across the microplate to account for any potential positional error. These internal controls allow us to identify the variants that are better than the parent with a higher confidence. Important to note, we use the term 'parent' instead of 'wildtype'

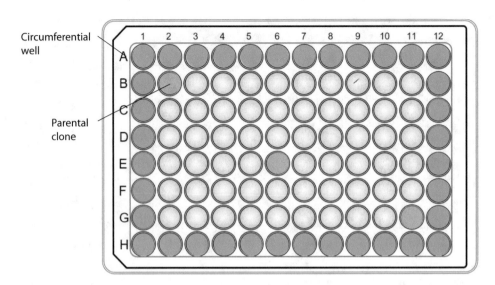

Fig. 9.1 A 96-well microplate. The circumferential wells are coloured in grey. Wells B2, E6 and G11 are inoculated with parental clones, and they serve as the internal controls for each microplate.

for internal control, because it should always be the parent from which the mutant library is created from. In ▶ Sect. 9.4, we will demonstrate analysis of the measurement data from each microplate. The values of the three internal controls are integral to this analysis.

9.1.3 **High-Throughput Cultivation**

Microbial cultivation in 96-well microplates is vastly different from doing it in Erlenmeyer shake flasks due to changes in:
- Culture volume
- Headspace
- Design and operating parameters of an incubator shaker
- Mixing

9.1.3.1 **Cell Settlement**
During microbial cultivation in 96-well microplates, you may observe cells settling to the bottom of the plates, which is not ideal. This problem is caused by two factors, which relate to one another:
- Poor mixing
- Cell death

Orbital shaking is used in almost all incubator shakers. It is a simple and non-invasive way of mixing. The choice of appropriate operating parameters for orbital mixing, especially the mixing frequency n (in 1/min) and the amplitude d_0 (in mm), depends on:
- Microplate filling volume V_F (in L)
- Well geometry, such as its diameter D_W (in mm) and height h

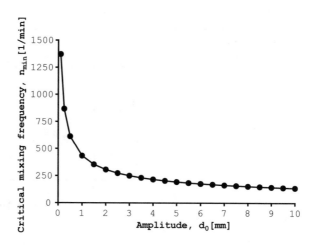

Fig. 9.2 Correlation between amplitude (d_0) and critical mixing frequency (n_{min}), calculated using σ of 72.7 mN/m (for water), D_w of 6.35 mm (for a standard 96-well microplate), V_F of 0.0002 L (recommended working volume of a standard 96-well microplate) and ρ of 0.977 kg/L (for water).

— Surface tension of the fluid (in mN/m)
— Fluid density ρ (in kg/L)
— Kinematic viscosity of the fluid v

The most important requirement for effective mixing is the formation of a macroscopic flow. As the microplate well volume decreases, the impact of surface tension increases. For this reason, it is necessary to generate a high centrifugal acceleration to achieve an intensive macroscopic flow. The force required for surface enlargement must be delivered by the centrifugal force. Centrifugal force exceeds surface tension when the shaking frequency is at or above the critical shaking frequency n_{min} (in 1/min), which can be described by the equation below (Fig. 9.2):

$$n_{min} = \sqrt{\frac{\sigma D_w}{4\pi V_F \rho d_0}}$$

The correlation in Fig. 9.2 tells us shaking incubators designed with smaller amplitude will require higher shaking frequency to ensure effective mixing. Poor mixing plus a small headspace in each well leads to poor aeration during cultivation, a factor that contributes to cell death. For aerobic cultivation, we recommend keeping the filling volume between 150 and 200 µL in the microplate and using a shaking frequency ≥ n_{min}.

9.1.3.2 Temperature

A high degree of temperature uniformity is important during cultivation. This ensures homogeneous cell growth condition for every variant across the plate, regardless of its positioning in the shaking incubator. Instrumentation is a major factor in temperature uniformity, thus an important consideration when purchasing a microplate shaking incubator. Further, cultivating at a temperature closer to the room temperature, *e.g.*, 30 °C instead of 37 °C for *E. coli*, will also reduce temperature differences.

9.1.3.3 Evaporation

In small volume cultivation, any liquid loss to evaporation will significantly change the total volume. Evaporation is a major issue when working with microplates. The circumferential wells display increased evaporation rate compared to the centrally

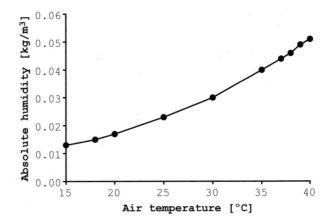

Fig. 9.3 Correlation between air temperature and absolute humidity.

located wells (■ Fig. 9.1). As a consequence, the measurement readings from these circumferential wells almost always show higher deviation or variability, compared to the rest of the wells. This phenomenon is referred to as the edge effect.

There are a few tips to reduce the edge effect:

- During clone picking, some researchers avoid using the circumferential wells, although they are filled with same volume of growth medium.
- Use low evaporation microplate lid and seal the microplate with parafilm.
- Use sealing foil designed to reduce evaporation.
- Humidify the incubator shaker during microbial cultivation, which is very effective in reducing evaporation. However, it is important to note the correlation between absolute humidity and air temperature as shown in ■ Fig. 9.3, which show the mass of water vapour that can be present in the air at the specific temperature.
- Reduce the microbial cultivation temperature.

9.1.4 Cell Disruption

If your target protein is located intracellularly, you would often need to disrupt or lyse the cells to release the target protein prior to assay. The process of cell disruption introduces numerous steps (*e.g.*, centrifugation, resuspension) into the HTS process. This adds to the time and resource requirement, but more importantly each additional step can contribute to errors during screening. In our laboratories, we tend to bypass this tedious step through the use of the Bacterial Extracellular Protein Secretion System (BENNY). BENNY utilises the *E. coli* osmotically-inducible protein Y (OsmY) as a protein fusion partner to direct the extracellular secretion of a target protein into the culture medium (see ▶ Sect. 4.2.4.2). In line with the KISS principle introduced earlier, BENNY keeps the process simple and straightforward by omitting cell disruption, nucleic acid degradation and cell debris centrifugation steps that will be discussed in this section.

9.1.4.1 Methods for Cell Disruption

If you need to carry out cell disruption, ■ Table 9.2 lists multiple ways that are compatible with the 96-well format. Often, methods are combined (*e.g.*, physical approach + enzymatic approach) to achieve a higher efficiency.

◼ **Table 9.2** Cell disruption techniques compatible with a 96-well format

Principle	Examples
Physical	Freeze & thaw
Enzyme	Lysozyme, rLysozyme, lysostaphin
Detergent	BugBuster, B-PER
Detergent and enzyme blend	CelLytic
Other chemicals	Polymyxin B

9.1.4.2 Viscosity

When bacterial cells are disrupted, the viscosity of the cell lysate increases due to the release of nucleic acid. Depending on the sample, it can sometimes be difficult to pipette or transfer the cell lysate accurately. If necessary, nucleases can be supplemented into the cell lysis buffer (*e.g.*, Benzonase, TurboNuclease, Pierce Universal Nuclease, and DNase I *etc*) to reduce viscosity. The use of nucleases inevitably adds to the screening cost.

9.1.4.3 Plate Centrifugation

After cell lysis and nucleic acid degradation, a plate centrifugation step is necessary to separate the cell lysate from the cell debris. This is a rate limiting step because most standard bench-top centrifuge can only accommodate two to eight microplates at once. Plate centrifugation is best done using V-bottom plates since it creates a more compact cell pellet, making it easier to aspirate the cell lysate.

9.1.5 Spectroscopic Measurement

Whenever you use a piece of equipment, it is advisable to first read its manual to understand its measurement principle, key features, and limitations.

9.1.5.1 Microplate Reader

When using a microplate reader, remember to check its measurement range and linearity range. For absorbance readers, the measurement range is usually 0–4 Abs and the linear range 0–2 Abs. Your measured value should fall within the measurement range and the linearity range (*e.g.*, below 2 Abs). Most microplate readers cannot reliably measure absorbance values above 3 Abs. Therefore. if your measured value is above the range, reduce the reaction time (▶ Sect. 9.2.2) or the enzyme concentration (Sect. ▶ 9.2.3).

9.1.5.2 Bubble

Protein solutions have foaming property. Try to avoid foam or bubble during pipetting. Bubbles in the assay plates will cause light scattering and erroneous signals. Briefly centrifuge the microplate or pop the few large bubbles with a pipette tip, before spectroscopic measurement.

9.1.5.3 Assay Signal Stability

When conducting a fluorescence assay, the photochemical destruction of the fluoro-phore is frequently observed (*i.e.*, photobleaching). Useful tips for minimizing pho-tobleaching include:
- Protect the sample from light using aluminium foil to minimize the sample's expo-sure to excitation illumination.
- Do not read the microplate twice.

9.2 Assay Conditions for Screening

The phrase "you get what you screen for" most appropriately describes this section. The assay condition is a key determinant of the screening outcome. We have learned in ▶ Chap. 7 about the chemistry, detection mode and format of an assay, as well as the parameters used to describe assay and enzyme performances. So how do we opti-mise an assay for screening purposes? There are two key questions here:
- What is the linear range of my assay?
- What is the enzyme property I want to improve?

The product linear detection range (see ▶ Fig. 7.11) defines the operation range for comparative work (*e.g.*, during screening). Within this range, the assay signal (*e.g.*, absorbance) is directly proportional to the analyte concentration (*e.g.*, enzymatic reaction product). It is important during screening, that the parental clone generate a signal that is in the lower half of the linear range to allow identification of at least a two-fold improved variant. Using ▶ Fig. 7.11 as an example, if the parental clone generates [Analyte] = 5 during screening, two-fold and three-fold improved variants with [Analyte] = 10 and 15, respectively, will have proportionately increased response signals that still lie within the linear range. These variants will be identified dur-ing screening. On the contrary, if the parental clone generates [Analyte] = 15 under screening conditions, a two-fold improved variant that generates [Analyte] = 30 will not be identified as it will give a response signal that is just slightly higher and out-side the linear range. The signal from the parental clone can most easily be adjusted by selecting a suitable reaction time (▶ Sect. 9.2.2) and enzyme concentration (▶ Sect. 9.2.3).

Since the analyte is a product of enzymatic reaction, a higher signal represents more reaction product. During screening, this can mean the variant has a higher enzyme velocity (*i.e.*, higher v) or more enzyme was expressed by the variant clone (*i.e.*, higher $[E]_0$) or both. When the improvement is due to a higher enzyme velocity, this can be attributed to a higher V_{max} or a lower K_m (refer to ▶ Sect. 7.2.3 for discus-sion on Michaelis-Menten kinetics). Whether it is a variant with higher V_{max} or lower K_m may seem like a random event with equal probability – that is not totally right. You can design the assay conditions to favour the specific property you seek.

Let's illustrate this using the substrate concentration in an assay. ▢ Figure 9.4 shows plots of the Michaelis-Menten kinetics relationship for a parental clone with K_m = 10 mM and V_{max} = 100 $\mu M.s^{-1}$ and improved variants that can be identified during screening. To isolate a variant with a higher V_{max}, we recommend using a substrate concentration that is 5–10 times the value of K_m. For instance, if a 80 mM substrate concentration is used during screening, improved variants will always have

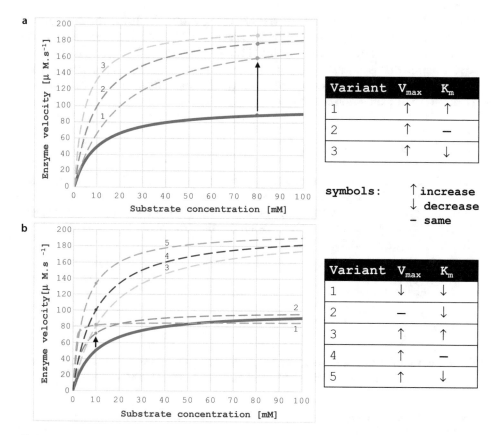

Fig. 9.4 Plots showing the Michaelis-Menten kinetics of a parental enzyme (blue solid line) and improved variants identified (dotted lines) during HTS. Substrate concentrations of 80 mM **a** and 10 mM **b** are used for the assay during HTS.

a higher V_{max}, despite the fact that K_m can remain the same, increase or decrease (**Fig. 9.4a**). Under a different condition where a substrate concentration at K_m or lower is used during screening, there are five potential outcomes for the improved variants identified (**Fig. 9.4b**). Among these five outcomes, three resulted in a lower K_m, one resulted in an unchanged K_m and one resulted in a higher K_m. There is also only a 60% probability that the variant has a higher V_{max}.

You have now seen examples of how assay conditions can affect the outcome of HTS. As much as it is possible, it is best to screen the enzyme under the conditions that you intend to use it. Let's look at the considerations when designing an assay for HTS.

9.2.1 Substrate Concentration

Assays should be conducted under the condition where <10% of substrate is converted. This ensures that we are measuring the initial velocity, where the enzyme velocity does not change with time. The rate of an enzymatic reaction is likely to fall when >10% of the substrate is converted. Keeping within the 10% substrate conversion guideline will ensure that the assay measures within the linear range of assay signal *vs* enzyme concentration during HTS. As such, we recommend a substrate concentration that is at least 10× the product concentration at the upper limit of your analyte linear detection range. Solubility, cost and availability are also considerations when choosing a substrate concentration.

9.2.2 Reaction Time

Controlling the reaction time is one simple way to ensure operation within the product linear detection range. To illustrate this, ◻ Fig. 9.5 shows the reaction progress curves of trypsin-catalysed release of *p*-nitroaniline from a trypsin pseudosubstrate N-CBZ-Gly-Pro-Arg-*p*-nitroanilide, assuming the hydrolytic reaction obeys the Michaelis-Menten model of

$$E + S \underset{k_{-1}}{\overset{k_1}{\Longleftrightarrow}} ES \overset{k_2}{\to} E + P$$

$$K_m = \frac{k_{-1} + k_2}{k_1}$$

The curves represent three separate reactions using the same substrate concentration of 100 μM, but with varied enzyme concentrations (10 nM in blue, 0.5 nM in

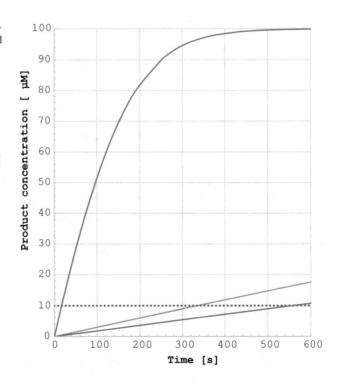

◻ **Fig. 9.5** The reaction progress curves of trypsin-catalysed release of *p*-nitroaniline from a trypsin pseudosubstrate N-CBZ-Gly-Pro-Arg-*p*-nitroanilide, assuming a Michaelis-Menten model. The three curves are obtained using a substrate concentration of 100 μM, but at varied enzyme concentrations (10 nM in blue, 0.5 nM in orange, and 0.3 nM in magenta). The reaction rate constants are $k_1 = 1.42$ μM^{-1} s^{-1}, $k_{-1} = 6.78$ s^{-1}, and $k_2 = 114$ s^{-1}. The red dotted line shows 10% substrate conversion.

orange, and 0.3 nM in magenta). The reaction rate constants are $k_1 = 1.42 \ \mu M^{-1} \ s^{-1}$, $k_{-1} = 6.78 \ s^{-1}$, and $k_2 = 114 \ s^{-1}$. The relationship between product concentration P and time t is described by the equations

$$\frac{dP}{dt} = \frac{k_2 E \left(S_0 - P\right)}{K_m + S_0 - P}$$

$$t = \frac{P}{k_2 E} + \frac{K_m}{k_2 E} \ln \frac{S_0}{S_0 - P}$$

where E = enzyme concentration, S_0 = initial substrate concentration, and K_m = Michaelis constant.

In this example, in order to keep substrate conversion below ~10% (red dotted line in ◻ Fig. 9.5), reactions should be stopped after 20 s (E = 10 nM), 5 min (E = 0.5 nM), and 9 min (E = 0.3 nM). During HTS, we recommend avoiding very short reaction time (*e.g.*, <2 min) because a slight delay in stopping or measuring the reaction will lead to significant variation in activity calculation. It is also important to ensure that reagents stored in a fridge or freezer have equilibrated to the assay temperature before use. Temperature variation can affect enzymatic activity, and this is particularly significant when using a short assay time.

9.2.3 Enzyme Concentration

Adjusting the enzyme concentration is another easy way to make sure we are operating within the product linear detection range. In the example presented in ◻ Fig. 9.5, we would choose an enzyme concentration of 0.3 nM, as it gives us a longer reaction time of 9 min. Apparent from ◻ Fig. 9.5, a lower enzyme concentration (*e.g.*, 0.5 nM or 0.3 nM) would allow us to capture the initial reaction rate. At high enzyme concentration (10 nM), the linearity of the reaction progress curve is transient, which does not allow a reliable deduction of initial reaction rate and comparison between variants and the parental clones. Equally important to bear in mind, enzyme tends to lose its activity quicker, when it is very diluted. As such, you would need to find the right trade-off.

9.2.4 Reaction Termination

An acid or a base is commonly used to terminate an enzymatic reaction. To keep screening simple, a termination reaction is only introduced when it is necessary or when it benefits the HTS process. Depending on the individual assay, the reasons for having a termination reaction can include:

- It is necessary for product formation.
- It enhances the signal intensity by a change in the product protonation state.
- It stabilizes the signal via a chemical reaction.

9.3 How Good Is Your HTS System?

A good understanding of the practical setup and assay conditions helps you to refine your HTS system. But to what extent do you need to optimise your HTS system? How would you know that your HTS system is good enough to be applied for screening your mutant library?

9.3.1 Background Plate Versus Wildtype Plate

Before you rush into library screening, it is advisable to first check the assay background and the precision of your HTS system, using the scheme outlined in ◘ Fig. 9.6. In this scheme, two microplates are prepared. The entire background plate is inoculated with your expression host harbouring an empty plasmid [*e.g.*, BL21(DE3) transformed with pET-19b], while the entire wildtype plate is inoculated with the same expression host but carrying plasmid with a target gene [*e.g.*, BL21(DE3) transformed with pET-19b-Tfu_0901]. By conducting a parallel cultivation, protein expression, protein extraction and enzymatic assays, we are able to obtain the assay background from the background plate, and the precision of the HTS system from the wildtype plate.

9

9.3.2 Assessment of HTS Data

Data processing is inevitable in HTS. Typical data output from a microplate reader is shown in ◘ Fig. 9.7, which is difficult to analyse and visualise. The operating software of most microplate readers allow you to export your data in .xlsx or .csv file

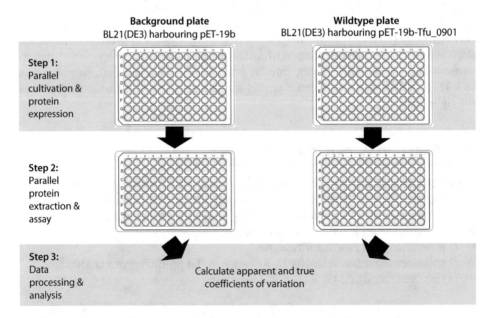

◘ **Fig. 9.6**   Procedures for checking assay background and precision of HTS system.

a   Background plate

	1	2	3	4	5	6	7	8	9	10	11	12
A	0.2645	0.2579	0.2589	0.2626	0.2580	0.2461	0.2515	0.2497	0.2577	0.2554	0.2600	0.2697
B	0.2574	0.2462	0.2407	0.2435	0.2432	0.2370	0.2383	0.2419	0.2349	0.2370	0.2269	0.2549
C	0.2445	0.2401	0.2388	0.2362	0.2370	0.2358	0.2323	0.2413	0.2273	0.2357	0.2380	0.2486
D	0.2424	0.2413	0.2386	0.2359	0.2429	0.2313	0.2301	0.2423	0.2317	0.2381	0.2390	0.2425
E	0.2554	0.2436	0.2478	0.2334	0.2353	0.2354	0.2394	0.2337	0.2403	0.2309	0.2355	0.2405
F	0.2556	0.2449	0.2360	0.2341	0.2294	0.2344	0.2436	0.2309	0.2387	0.2286	0.2390	0.2410
G	0.2501	0.2431	0.2313	0.2396	0.2366	0.2303	0.2383	0.2329	0.2387	0.2279	0.2358	0.2477
H	0.2668	0.2602	0.2490	0.2522	0.2478	0.2421	0.2488	0.2437	0.2541	0.2440	0.2493	0.2565

	Entire plate
Mean	0.2427

b   wildtype plate

	1	2	3	4	5	6	7	8	9	10	11	12
A	0.4941	0.4143	0.4336	0.4043	0.3926	0.3632	0.3644	0.3880	0.3874	0.3871	0.3861	0.4197
B	0.4470	0.3887	0.3661	0.3778	0.3627	0.3658	0.3555	0.3653	0.3776	0.3642	0.3597	0.3794
C	0.4065	0.3735	0.3741	0.3764	0.3786	0.3561	0.3573	0.3585	0.3567	0.3651	0.3487	0.3668
D	0.4260	0.3788	0.3745	0.3758	0.3532	0.3449	0.3349	0.3554	0.3384	0.3517	0.3533	0.3564
E	0.4185	0.3777	0.3765	0.3762	0.3691	0.3669	0.3803	0.3711	0.3706	0.3617	0.3986	0.3826
F	0.4381	0.3787	0.3795	0.3884	0.3689	0.3828	0.3661	0.3945	0.3754	0.3699	0.3576	0.3839
G	0.4830	0.3830	0.4057	0.4089	0.3797	0.3575	0.3654	0.3861	0.3822	0.3677	0.3671	0.3824
H	0.524	0.4593	0.4169	0.4234	0.4047	0.4202	0.4174	0.4018	0.4085	0.3790	0.4175	0.4024

	Entire plate	Circumferential wells removed
Mean	0.3852	0.3701
SD	0.0319	0.0146
CV	8%	4%

c   Wildtype – Background

	1	2	3	4	5	6	7	8	9	10	11	12
A	0.2514	0.1716	0.1909	0.1616	0.1499	0.1205	0.1217	0.1453	0.1447	0.1444	0.1434	0.1770
B	0.2043	0.1460	0.1234	0.1351	0.1200	0.1231	0.1128	0.1226	0.1349	0.1215	0.1170	0.1367
C	0.1638	0.1308	0.1314	0.1337	0.1359	0.1134	0.1146	0.1158	0.1140	0.1224	0.1060	0.1241
D	0.1833	0.1361	0.1318	0.1331	0.1105	0.1022	0.0922	0.1127	0.0957	0.1090	0.1106	0.1137
E	0.1758	0.1350	0.1338	0.1335	0.1264	0.1242	0.1376	0.1284	0.1279	0.1190	0.1559	0.1399
F	0.1954	0.1360	0.1368	0.1457	0.1262	0.1401	0.1234	0.1518	0.1327	0.1272	0.1149	0.1412
G	0.2403	0.1403	0.1630	0.1662	0.1370	0.1148	0.1227	0.1434	0.1395	0.1250	0.1244	0.1397
H	0.2814	0.2166	0.1742	0.1807	0.1620	0.1775	0.1747	0.1591	0.1658	0.1363	0.1748	0.1597

	Entire plate	Circumferential wells removed
Mean	0.1425	0.1273
SD	0.0319	0.0146
CV	22%	11%

Fig. 9.7   Analysis of HTS data.

formats, which means you can open and process your data file using Microsoft Excel. Key parameters that are used to assess the precision of your HTS is shown in the following sections.

9.3.2.1 Mean

Mean is also known as the average. It can be calculated using the formula of:

$$Mean = \mu = \frac{\sum_{i=1}^{N} x_i}{N}$$

If you open your data file in Excel, the easiest way to calculate mean is to apply the AVERAGE function in Excel.

9.3.2.2 Standard Deviation (SD)

Standard deviation (SD) measures the spread of the data about the mean value. It is calculated using the formulae below:

$$Variance = \sigma^2 = \frac{\sum_{i=1}^{N} (x_i - \mu)^2}{N-1}$$

$$Standard\ deviation\,(SD) = \sigma = \sqrt{\frac{\sum_{i=1}^{N} (x_i - \mu)^2}{N-1}}$$

Alternatively, you can apply the STDEV function in Excel.

9.3.2.3 Coefficient of Variation (CV)

Coefficient of variation (CV), similar to SD, is a measure of variability:

$$Coefficient\ of\ variation\,(CV) = \frac{\sigma}{\mu} \times 100\%$$

9.3.2.4 Apparent and True Coefficients of Variation

Using the example in ◻ Fig. 9.7, the assay background is the mean value obtained from the background plate (0.2427, calculated using dataset A). If we turn our attention to the wildtype plate, the mean value is 0.3852 (calculated using dataset B). This means that an increase of 0.1425 in assay value (0.3852–0.2427 = 0.1425) is due to the enzymatic reaction. If we subtract the background (0.2427) from each value in the wildtype plate, we obtain a new set of values, dataset C, where C = B – A. The CV of dataset B is 8%, which is the apparent coefficient of variation. The true coefficient of variation is the CV of dataset C, which is 22%. The high CVs of B and C are mainly contributed by the edge effect. If we remove those values from all the circumferential wells, the CVs of B and C are now reduced to 4% and 11%, respectively. Typically, a true CV of <15% is considered acceptable for library screening.

9.4 Analysing and Presenting Library Screening Result

In this section, we will guide you through analysing and presenting your library screening data, using the library plate shown in ◘ Fig. 9.8.

9.4.1 Internal Controls

The three internal controls in a microplate are the parental clones (◘ Fig. 9.8, coloured in red). They serve as the benchmark for enzyme performance comparison. As such, it is imperative to first analyse these three values by calculating their mean, standard deviation and coefficient of variation, as illustrated below.

$$Mean = \frac{2.037 + 1.994 + 2.103}{3} = 2.045$$

$$Standard\ deviation\,(SD) = \sqrt{\frac{(2.037 - 2.045)^2 + (1.994 - 2.045)^2 + (2.103 - 2.045)^2}{3 - 1}}$$

$$= 0.055$$

$$CV = \frac{0.055}{2.045} \times 100\% = 2.69\%$$

	1	2	3	4	5	6	7	8	9	10	11	12
A	3.099	0.322	2.509	0.276	0.273	0.876	0.636	0.857	0.294	2.042	2.129	3.192
B	1.354	2.037	2.263	0.283	1.636	2.345	0.680	1.429	0.283	0.248	0.386	1.891
C	2.453	0.298	0.325	0.269	1.641	0.271	0.807	0.297	0.313	0.272	0.264	1.514
D	0.431	0.432	0.283	0.367	2.035	0.288	2.252	0.425	0.608	1.670	0.303	2.184
E	0.876	0.291	0.347	0.272	0.276	1.994	0.436	0.253	0.270	1.867	2.528	0.281
F	3.777	0.429	0.370	1.549	0.263	0.276	0.266	1.865	1.703	0.495	0.456	0.303
G	0.494	2.345	3.318	0.349	0.343	0.272	0.340	0.328	0.466	0.293	2.103	0.653
H	3.644	3.163	0.328	0.446	3.467	0.346	3.232	0.645	3.263	0.336	0.418	3.271

◘ **Fig. 9.8** Typical data output from a microplate reader for a library plate. The values of the three internal controls are coloured in red.

9.4.2 Data Presentation

For data presentation, we recommend two formats:
- Apply a conditional formatting on the cells (■ Fig. 9.9): This format allows you to 'highlight' clones that are better than the parental clones and identify the best clones. You might have notice that, in ■ Fig. 9.7, we have applied conditional formatting to make it easier for you to spot the edge effect.
- Plot the data in descending order of assay values (■ Fig. 9.10): This format makes it easier to evaluate the quality of your mutant library. If most of the clones are inactive, this would mean the mutation frequency of your mutant library is too high.

	1	2	3	4	5	6	7	8	9	10	11	12
A	3.099	0.322	2.509	0.276	0.273	0.876	0.636	0.857	0.294	2.042	2.129	3.192
B	1.354	2.037	2.263	0.283	1.636	2.345	0.680	1.429	0.283	0.248	0.386	1.891
C	2.453	0.298	0.325	0.269	1.641	0.271	0.807	0.297	0.313	0.272	0.264	1.514
D	0.431	0.432	0.283	0.367	2.035	0.288	2.252	0.425	0.608	1.670	0.303	2.184
E	0.876	0.291	0.347	0.272	0.276	1.994	0.436	0.253	0.270	1.867	2.528	0.281
F	3.777	0.429	0.370	1.549	0.263	0.276	0.266	1.865	1.703	0.495	0.456	0.303
G	0.494	2.345	3.318	0.349	0.343	0.272	0.340	0.328	0.466	0.293	2.103	0.653
H	3.644	3.163	0.328	0.446	3.467	0.346	3.232	0.645	3.263	0.336	0.418	3.271

■ **Fig. 9.9** Conditional formatting is applied to the data set by colouring the cells in graded colour scale, with largest value in dark grey and smallest value in white.

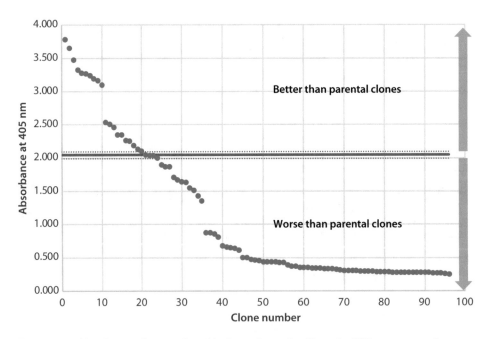

■ **Fig. 9.10** Absorbance values are plotted in descending order. The red solid line represents the mean value (μ) of the three internal controls. The two red dotted lines represent mean ± standard deviation ($\mu \pm \sigma = 2.045 \pm 0.055$). In term of enzyme performance, clones above the upper dotted line are better than the parental clones. Clones below the lower dotted line are worse than the parental clones.

When analysing HTS data, it is again important to look out for potential edge effect. In the example shown in ◻ Fig. 9.9, there is a sizable number of 'better' clones in the circumferential wells. This could potentially be due to evaporation, leading to an apparent higher protein concentration in those wells.

9.5 Re-screening

No screening is perfect and error-free. The possibilities of picking a false positive and missing out on a false negative are real. Our primary step is to develop a precise and robust HTS system to reduce such errors. Our second recommendation is to re-screen a few clones, before using it in a next round of mutagenesis and screening. ◻ Figure 9.11 shows our typical scheme of re-screening:
- From the assay plate, identify the improved protein variants.
- From the corresponding master plate, pick the clones expressing those improved protein variants and prepare a tube culture for each clone.
- From the tube culture, isolate the plasmid DNA. The DNA extracted serves two purposes, protein expression and DNA sequencing.
- Transform the plasmid DNA into the same expression host used in HTS.
- Pick a single colony from the agar plate and prepare a tube culture.
- Do a small-scale protein expression in Erlenmeyer shake flask (25–50 mL).
- Extract protein and re-do the assay in a microplate or a cuvette (for larger reaction volume).

9.6 Adjusting Your HTS System as You Go Along

As your directed evolution progresses, your protein variants become better, giving you higher assay signal value as depicted in ◻ Fig. 9.12. Eventually, you will reach the limit for reliable measurement of your microplate reader (typically 3 Abs) or exceed the product linear detection range. Therefore, your HTS system must adapt. You can adjust your HTS screening by:

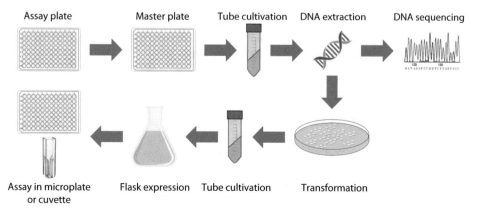

Assay plate   Master plate   Tube cultivation   DNA extraction   DNA sequencing

Assay in microplate   Flask expression   Tube cultivation   Transformation
or cuvette

◻ **Fig. 9.11**   Re-screening scheme.

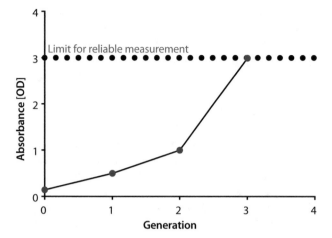

Fig. 9.12 Incremental enhancement of enzyme performance through directed evolution.

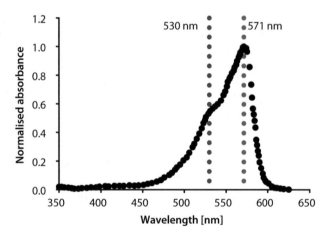

Fig. 9.13 Absorption spectrum of a dye and its absorption maximum.

9

- Reducing the reaction time.
- Lowering the volume of cell lysate used.
- Use another wavelength to reduce the sensitivity of your assay. In ◘ Fig. 9.13, the absorption spectrum of a dye is shown, with an absorption maximum at 571 nm. The assay is most sensitive when measured at this wavelength. To reduce the sensitivity, one can use a wavelength such as 530 nm.

Take-Home Messages

1. Screening is a sifting process that involves the use of an assay.
2. When screening is conducted in a high-throughput manner, it is more likely to run into technical challenges due to parallelization and miniaturization of the assay.
3. BENNY streamlines the directed evolution workflow, by secreting the enzyme involved extracellularly thereby bypassing the cell lysis step.

4. In a microplate screening, internal controls serve as the benchmark for comparing parental clones and protein variants.
5. Evaporation from microplates causes the edge effect observed in a microplate assay.
6. An enzymatic assay should be conducted in a condition whereby only less than 10% of the substrate is converted.
7. The types of mutant obtained is highly dependent on how the mutant library is screened (*i.e.*, you get what you screen for).
8. The true coefficient of variation provides a good indication of the reliability of a high-throughput screening system.
9. A high-throughput screening system must adapt as the protein variants improve.

Exercise

(a) Calculate the mean, standard deviation and coefficient of variation of the three internal controls from library plate A (◻ Fig. 9.14).
(b) Repeat the calculations for library plate B.
(c) What information could you deduce from the calculations in (a) and (b)?
(d) Do you observe edge effect in the library plates? Justify your answer.
(e) Which library plate is prepared using a higher mutation rate? Justify your answer.

Library plate A

Value	1	2	3	4	5	6	7	8	9	10	11	12
A	3.0168	2.7325	0.3084	2.6126	2.7281	1.5628	0.2965	2.6557	0.3031	1.6728	2.9854	0.3397
B	2.8321	1.5489	0.2835	2.4599	0.6319	2.1593	2.0365	2.3203	2.2247	2.0727	2.4478	1.3952
C	0.3220	0.4794	2.0434	2.0583	1.5869	0.3944	1.8808	2.1320	0.2936	0.2975	2.4203	2.4840
D	0.2876	2.4027	2.2735	2.2956	2.1808	0.3024	0.4016	1.9612	1.6291	1.6010	0.4093	2.5603
E	0.3518	1.1025	2.1261	2.1620	0.6677	1.9284	2.1531	2.0346	1.7136	0.2782	1.9859	2.5228
F	2.2177	2.4239	2.6022	2.2113	0.2962	2.1261	0.2728	0.4398	1.9383	1.8842	0.3201	2.5759
G	2.5048	0.3934	2.3427	2.4689	0.2948	2.3743	2.2429	2.1452	2.4623	2.3412	2.4945	2.6508
H	0.3067	0.4407	2.7501	2.0792	2.4729	2.4550	0.3951	0.2801	2.2500	2.5603	2.7094	2.1352

Library plate B

Value	1	2	3	4	5	6	7	8	9	10	11	12
A	3.0989	0.3218	2.5087	0.2763	0.2729	0.8756	0.6360	0.8572	0.2939	2.0422	2.1291	3.1921
B	1.3540	1.9983	2.2634	0.2825	1.6362	2.3447	0.6796	1.4289	0.2832	0.2479	0.3864	1.8909
C	2.4528	0.2976	0.3250	0.2685	1.6409	0.2706	0.8066	0.2974	0.3133	0.2720	0.2639	1.5140
D	0.4309	0.4324	0.2825	0.3668	2.0345	0.2877	2.2517	0.4252	0.6076	1.6695	0.3025	2.1835
E	0.8755	0.2905	0.3471	0.2715	0.2760	1.9823	0.4361	0.2531	0.2702	1.8669	2.5279	0.2813
F	3.7770	0.4291	0.3700	1.5488	0.2625	0.2759	0.2663	1.8647	1.7033	0.4947	0.4564	0.3032
G	0.4936	2.3448	3.3184	0.3492	0.3427	0.2716	0.3400	0.3280	0.4657	0.2926	1.9920	0.6533
H	3.6444	3.1630	0.3281	0.4461	3.4670	0.3463	3.2322	0.6446	3.2634	0.3362	0.4180	3.2709

◻ **Fig. 9.14** Assay data from two library plates **A** and **B**. The values for internal controls are coloured in red.

Further Reading

Hermann R, Lehmann M, Buchs J (2003) Characterization of gas-liquid mass transfer phenomena in microtiter plates. Biotechnol Bioeng 81(2):178–186

Kong F, Yuan L, Zheng YF, Chen W (2012) Automatic liquid handling for life science: a critical review of the current state of the art. J Lab Autom 17(3):169–185

Trask OJ (2018) Guidelines for microplate selection in high content imaging. Methods Mol Biol 1683:75–88

Protein Variants Analysis and Characterization

Contents

© Springer Nature Switzerland AG 2020
T. S. Wong, K. L. Tee, *A Practical Guide to Protein Engineering*, Learning Materials in Biosciences,
https://doi.org/10.1007/978-3-030-56898-6_10

What You Will Learn in This Chapter

In this chapter, we will learn to:
- prepare samples for DNA sequencing.
- analyse sequencing chromatogram and identify mutation(s).
- map mutation(s) onto 3D structure of protein.
- understand the difference between additive interaction and epistatic interaction between mutations.
- design a protein purification scheme.
- quantify purified protein using molar extinction coefficient.
- quantify purified protein using Bradford assay or BCA assay.
- obtain initial reaction rate and enzyme kinetic parameters.
- use the Solver in Microsoft Excel for non-linear regression.

Analysis and characterization of an improved protein variant is an exciting part of protein engineering, as you are one step closer to:
- Your thesis, if you are a student doing a research project
- A research publication or a grant application, if you are a research scientist
- A patent, if you are an entrepreneur interested in commercialisation and licensing
- Delivering a product to your client, if you run a protein engineering service company
- A new product or a new biocatalytic process, if you are an application scientist

10

Whichever role you are in, you are expected to describe your improved protein variant quantitatively and compare it against the wildtype protein. This chapter is dedicated to answering two important questions:
- Where and what are the mutations that confer the improved phenotype?
- How much better is the protein variant?

10.1 Mutation

Let's begin with identifying the mutation! You will need to apply the material you've learned on nucleotide substitution and amino acid substitution in ▶ Sect. 8.1.

10.1.1 DNA Sequencing and Sequencing Chromatogram

We have briefly introduced DNA sequencing in ▶ Sect. 5.4, a method to verify a clone after gene cloning. Shown in ▶ Fig. 9.11, DNA sequencing is also a direct way to determine the nature and the location of the mutation(s) in the gene encoding the improved protein variant.

10.1.1.1 Preparing DNA Sample for Sequencing

Preparing DNA sample for sequencing is straightforward. Plasmid DNA purified using standard plasmid prep kits (*e.g.*, Qiagen, Macherey-Nagel, Omega Bio-tek, Thermo Fisher Scientific, and New England Biolabs *etc*) are usually adequate for DNA sequencing. The recommended sample preparation guideline is:

- Template DNA concentration: 50–100 ng/μL
- Template volume: at least 15 μL for 1–4 sequencing reactions, or 20 μL for 5–8 reactions
- Sequencing primer concentration: 10 pmol/μL in double distilled water or 5 mM Tris-HCl (if you are using your own sequencing primer)
- Primer volume: at least 15 μL, and add 5 μL for every additional sequencing reaction
- Prepare template DNA and primer in 1.5-mL microcentrifuge tubes and label all tubes properly (*e.g.*, using pre-paid sequencing labels)
- Samples are sent by post or dropped at designated sample collection points

10.1.1.2 Interpretation of Sequencing Chromatogram

Nowadays the turnaround time of DNA sequencing is very short, typically 2–3 days. When you receive your DNA sequencing result, you will notice that there are three files per sequencing reaction:

- .ab1 (ABI sequencer data file): Known as the trace file, it includes raw data that is the output from Applied Biosystems' sequencing analysis software. .ab1 files include quality information about the base calls, the chromatogram (also called the electropherogram), and the DNA sequence. .ab1 files are invented a long time ago. They are not designed to store large sequence/chromatogram, and are occasionally corrupted.
- .scf (standard chromatogram format): Like .ab1 files, .scf files are also trace files that include quality information about the base calls, the chromatogram, and the DNA sequence. .scf files do not have the problems of .ab1 files. On top of that, they are designed to be easily compressible, which is important for storing large amount of chromatogram files. Most sequence assembly or viewing software are capable of reading both .ab1 and .scf files.
- .seq: Known as the sequence file, it is just a plain text file containing the DNA sequence.

All three file types can be opened using SnapGene Viewer, which was introduced in ▶ Chap. 2. The chromatogram itself tells us if the sequencing reaction is successful or has failed (◘ Table 10.1). A successful sequencing reaction gives a chromatogram that is characterised by defined and evenly spaced peaks with high signal-to-noise

◘ **Table 10.1** Successful *vs* failed sequencing reactions

Successful		Failed	
Chromatogram	**Why successful?**	**Chromatogram**	**Why failed?**
• Peaks well-formed and distinctive • Uniform peak separation • Absence of background signals	• Appropriate template and primer concentration • Good DNA purity • Optimum primer design • Good primer quality	• Absence of clearly defined peaks • Occurrence of excess dye peaks • Low signal-to-noise ratio	• Insufficient or poor quality template DNA and/or primer • Primer binding site absent, deleted or mutated

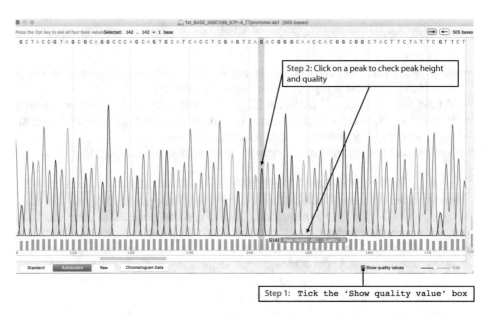

Step 2: Click on a peak to check peak height and quality

Step 1: Tick the 'Show quality value' box

☐ **Fig. 10.1** Chromatogram of a successful DNA sequencing reaction. The .ab1 file is opened using SnapGene Viewer.

10

ratio and a quality score of >40 (☐ Fig. 10.1). Sometimes, you might get a chromatogram with multiple sequence signals, in which peaks are not evenly spaced or overlapped or with artefacts beneath the peaks. This is an indication of contaminated or poor quality template DNA/primer or multiple priming events.

There are multiple ways of identifying mutations in a gene via sequence alignment:
- SnapGene: You can perform sequence alignment using the full version of SnapGene (not the free-of-charge SnapGene Viewer)
- BLAST and Clustal Omega: You have learned these tools in ▶ Sects. 2.2.1.3 and 2.2.1.4 for protein sequence analysis. They are also excellent tools for nucleotide sequence alignment.

There are a few important tips for you to consider when analysing DNA sequencing data:
- Expect to get 700–800 bases of reliable DNA sequence. This may vary with different sequencing service provider.
- The first 20–30 bases usually give poor and unresolved signals. Design your sequencing primer 50 bp upstream from the sequence of interest.
- Always check the chromatogram to confirm the presence of a mutation.
- Contact the sequencing provider for advice if you are confident that a failed sequencing reaction is not due to poor sample preparation.

10.1.2 **Structure-Function Relationships**

After sequencing data analysis, you will gather the information below:
- Nucleotide substitution

- Location of the nucleotide substitution within the gene sequence
- Amino acid substitution
- Location of the amino acid substitution within the protein sequence
- Location of the amino acid substitution within a protein structure (if available) or a protein model

This information will allow us to interrogate structure-function relationship, through answering the questions below:
- Is the mutation located on the protein surface?
- Is the mutation located in the active site? Does the side chain interact with the catalytically critical residues?
- Is the mutation located along the substrate channel? Does the side chain interact with the substrate?
- Is the mutation located at the oligomerization interface?
- Does the mutation affect cofactor binding?
- Does the mutation create additional bond(s)?

You can certainly apply your knowledge in protein structural analysis (▶ Chap. 3) to address the aforementioned questions. Some protein engineers would go a step further by conducting molecular simulations or attempting protein crystallography to gain molecular understanding of the mutation(s).

10.1.3 Additive Interaction Versus Epistatic Interaction

Your improved protein variant may have more than one mutation. In this case, there is a possibility that additive interaction or epistatic interaction exists among the mutations (◘ Fig. 10.2). Two mutations are considered to be purely additive (or multiplicative), if the effect of the double mutation is the sum of the effects of the single mutations. In other words, the effects of these uncoupled mutations are independent of one another. Epistasis, on the other hand, describes the phenomenon in which the effect of one mutation is dependent on the presence or absence of the other mutation. To investigate potential interactions between mutations, we would need to study these mutations singly and in combinations, which requires that we create a series of mutants using site-directed mutagenesis (▶ Sect. 8.4.2.2).

10.2 Protein Purification

Purified protein should be used for characterization, for the reasons listed below:
- To ascertain that the improved phenotype is attributed to the target protein.
- To have a more accurate quantification of protein concentration, which is important in calculating kinetic parameters, for example.
- To avoid potential interference from the host cell proteins or metabolites.
- To allow for other biophysical studies that will enhance our understanding of the target protein (*e.g.*, X-ray crystallography or NMR *etc*).

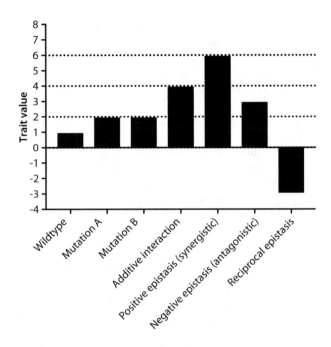

Fig. 10.2 Additive *vs* epistatic interactions between mutations.

10.2.1 **Protein Purification Methods**

Most protein purification methods are based on chromatography, exploiting various protein properties as summarised in ◻ Table 10.2. The first four methods in the table (AC, IMAC, IEX and GF/SEC) are most frequently used. More often than not, they are applied in combination to achieve a protein preparation with acceptable purity.

AC separates proteins on the basis of a reversible interaction between the target protein and a specific ligand attached to a chromatography matrix. The interaction can be biospecific, for example, antibodies binding to Protein A, or nonbiospecific, for example, a protein binding to maltose. IMAC, a subcategory of AC, is based on the interaction of proteins with exposed histidine residues with divalent metal ions (*e.g.*, Ni^{2+}, Cu^{2+}, Zn^{2+}, Co^{2+}) immobilized on matrix via a chelating ligand. IEX is based on the reversible interaction between a charged protein and an oppositely charged chromatography medium. GF is simple to use and allows separation of proteins with differences in molecular size, under mild conditions. HIC and RPC separate proteins with differences in hydrophobicity, with RPC requiring the use of organic solvents. Last but not least, CF separates proteins according to differences in their pI values.

Many chromatographic system and column suppliers [*e.g.*, Cytiva (formerly GE Healthcare Life Science), Bio-Rad, Thermo Fisher Scientific] produce comprehensive protein purification handbooks. These are excellent resources that we highly recommend to students who aim to acquire protein purification knowledge and skills:

☐ **Table 10.2** Protein purification methods

Protein purification method	Protein property exploited	Columns[a]
Affinity chromatography (AC)	Ligand recognition	GSTrap, MBPTrap, StrepTrap, strepavidin, heparin, Protein A, rProtein A, Protein G, benzamidine
Immobilized metal ion affinity chromatography (IMAC)	Metal ion binding	HisTrap
Ion exchange chromatography (IEX)	Protein charge	Q, SP, S, DEAE, CM
Gel filtration (GF) or size exclusion chromatography (SEC)	Protein size	Superdex 30, Superdex 75, Superdex 200
Hydrophobic interaction chromatography (HIC)	Protein hydrophobicity	Phenyl, Butyl, Octyl
Reverse phase chromatography (RPC)	Protein hydrophobicity	µRPC C2/C18, Source 5RPC, Source 15 RPC, Source 30RPC
Chromatofocusing	Protein isoelectric point (pI)	Mono P, PBE118, PBE94

[a]Columns from Cytiva (formerly GE Healthcare Life Sciences) are used as examples

- Cytiva's Principles & Methodology Handbooks (▶ https://www.cytivalifesciences.com/en/us/support/handbooks)
- Bio-Rad's Protein Purification Solutions (▶ https://info.bio-rad.com/PPD-RS-LP1?source_wt=proteinpurificationupdate_surl)
- Thermo Fisher Scientific's Protein Purification Support (▶ https://www.thermofisher.com/uk/en/home/technical-resources/technical-reference-library/protein-purification-isolation-support-center/protein-purification-support.html)

10.2.2 Protein Purification Strategy

To obtain a protein of sufficient quantity and purity, we advocate a three-stage purification: Capture, Intermediate Purification and Polishing (CIPP; ☐ Fig. 10.3). The objective of the capture stage is to isolate, concentrate, and stabilize the target product. After the capture stage, the product is concentrated and in an environment that will conserve its potency or activity. A significant removal of critical contaminants (*e.g.*, host cell proteins, nucleic acids, metabolites) can also be achieved. Impurity removal continues in the intermediate purification stage. If the capture step is efficient, the intermediate purification stage can be omitted. Finally, in

☐ Fig. 10.3 A 3-stage purification based on capture, intermediate purification and polishing (CIPP).

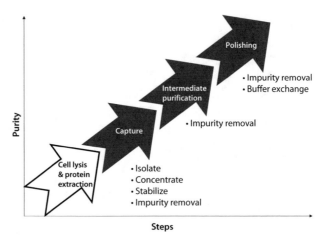

the polishing stage, impurities that exist in trace amounts are further removed. Importantly, the target protein is also transferred to a condition suitable for use or long-term storage.

To reinforce your knowledge on protein purification, let's design a scheme for the purification of Tfu_0901$_{37-466}$. We have already created the construct pET-19b-Tfu_0901 in ▶ Chap. 5 for recombinant protein expression in *E. coli*. The recombinant protein contains a 10×His tag and an enterokinase (EK) recognition site to facilitate protein purification (☐ Fig. 10.4). The first thing we ought to do is check the molecular weight (MW), isoelectric point (pI), extinction coefficient (ε) of the protein using ProtParam (▶ Sect. 2.2.1.1), pre- and post-EK cleavage:

- MW: This tells us the expected protein size, which is important when we run an SDS-PAGE during protein purification.
- pI: This informs our selection of a buffer system of the right pH for protein purification and protein storage.
- ε: This allows us to quantify protein concentration after purification.

Based on the pET-19b-Tfu_0901 construct, we propose the following purification scheme (☐ Fig. 10.4):

- Capture the His-tagged Tfu_0901$_{37-466}$ from the cell lysate using a HisTrap column.
- Carry out a desalting step to remove imidazole from the eluate.
- Add recombinant His-tagged enterokinase directly into the desalted sample to cleave the 10×His tag.
- Reload the sample onto a HisTrap column. The cleaved 10×His tag, the His-tagged enterokinase, and the uncleaved His-tagged Tfu_0901$_{37-466}$ will re-bind the HisTrap column, leaving the cleaved Tfu_0901$_{37-466}$ in the flow-through.
- Polish the protein sample by a gel filtration step using Superdex 200 column.

After a first attempt at this purification scheme, you would know which steps are working more effectively and which steps less effectively. To obtain a pure protein, we may need to optimise the washing and elution of each chromatographic step. If purity is an issue, we could add an additional IEX step using an anion exchanger (*e.g.*, Q or DEAE) before the GF step.

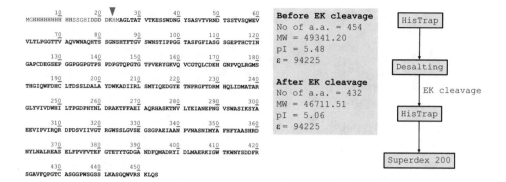

```
        10          20    ▼    30          40          50          60
MGHHHHHHHH HHSSGHIDDD DKHMAGLTAT VTKESSWDNG YSASVTVRND TSSTVSQWEV

        70          80          90         100         110         120
VLTLPGGTTV AQVWNAQHTS SGNSHTFTGV SWNSTIPPGG TASFGFIASG SGEPTHCTIN

       130         140         150         160         170         180
GAPCDEGSEP GGPGGPGTPS PDPGTQPGTG TPVERYGKVQ VCGTQLCDEH GNPVQLRGMS

       190         200         210         220         230         240
THGIQWFDHC LTDSSLDALA YDWKADIIRL SMYIQEDGYE TNPRGFTDRM HQLIDMATAR

       250         260         270         280         290         300
GLYVIVDWHI LTPGDPHYNL DRAKTFFAEI AQRHASKTNV LYEIANEPNG VSWASIKSYA

       310         320         330         340         350         360
EEVIPVIRQR DPDSVIIVGT RGWSSLGVSE GSGPAEIAAN PVNASNIMYA FHFYAASHRD

       370         380         390         400         410         420
NYLNALREAS ELFPVFVTEF GTETYTGDGA NDFQMADRYI DLMAERKIGW TKWNYSDDFR

       430         440         450
SGAVFQPGTC ASGGPWSGSS LKASGQWVRS KLQS
```

Before EK cleavage
No of a.a. = 454
MW = 49341.20
pI = 5.48
ε = 94225

After EK cleavage
No of a.a. = 432
MW = 46711.51
pI = 5.06
ε = 94225

HisTrap

Desalting

EK cleavage

HisTrap

Superdex 200

◻ **Fig. 10.4**   Protein purification scheme for Tfu_0901$_{37-466}$. His-tag is shown in bold red, enterokinase (EK) recognition site in bold blue, and Tfu_0901$_{37-466}$ in bold black. The red triangle indicates the position of EK cleavage. Protein parameters (MW and pI) are calculated using ProtParam. Protein purification scheme is designed using columns from Cytiva (formerly GE Healthcare Life Sciences).

In the proposed purification scheme, you will require His-tagged enterokinase. His-tagged enterokinase can be bought from commercial suppliers, but the production of recombinant enterokinase in *E. coli* is also well documented (PMIDs 24184090, 17516250, 12963350). With the knowledge and skills you have acquired so far, you should be able to produce recombinant enterokinase yourself.

What buffers should we use during purification? We have provided our suggestions in ◻ Table 10.3, which are based on the following facts:

- The pI values of the protein are 5.48 (pre-cleavage) and 5.06 (post-cleavage). Buffers with a pH of at least one unit away from the pI would be suitable.
- Enterokinase works best in Tris and MES, with pH between 7.0 and 8.0. High NaCl concentration or phosphate buffer reduces its activity. Ca^{2+} causes an activation of hydrolysis for some substrates.
- For long-term protein storage, storage buffer is supplemented with 10% (v/v) glycerol.

10.2.3 **Protein Quantification**

Immediately after protein purification, we recommend protein quantification. Regardless of the method you decide to use, protein quantification is only accurate when the protein is relatively pure. In this section, we will look at two common ways of protein quantification: (a) extinction coefficient and (b) protein assays.

10.2.3.1 **Extinction Coefficient**
The molar extinction coefficient of a protein is related to its tryptophan (W), tyrosine (Y) and cysteine (C) amino acid composition. We have taught you how to use ProtParam (▶ Sect. 2.2.1.1) to calculate this value. Alternatively, this value can be approximated using the following equation where nW, nY and nC refer to the number of W, Y and C residues within the protein sequence, respectively:

$$\varepsilon = (nW \times 5500) + (nY \times 1490) + (nC \times 125)$$

◻ **Table 10.3**   Buffers for the purification of Tfu_0901$_{37-466}$

Step	Buffer A	Buffer B	Note
1st HisTrap	50 mM sodium phosphate 300 mM NaCl 10 mM imidazole pH 8.0	50 mM sodium phosphate 300 mM NaCl 250 mM imidazole pH 8.0	Standard buffers for HisTrap
Desalting	20 mM Tris-HCl 50 mM NaCl 2 mM CaCl$_2$ pH 7.4		Reaction buffer for enterokinase
2nd HisTrap	50 mM sodium phosphate 300 mM NaCl 10 mM imidazole pH 8.0	50 mM sodium phosphate 300 mM NaCl 250 mM imidazole pH 8.0	Standard buffers for HisTrap
Gel filtration	25 mM Tris-HCl 150 mM NaCl 10% (v/v) glycerol pH 7.5		Standard protein storage buffer

10

To reiterate, the protein concentration (C) is calculated using the Lambert-Beer Law, where A and l are absorbance at 280 nm and pathlength, respectively:

$$C = \frac{A}{\varepsilon l}$$

10.2.3.2 Protein Assay

Several colorimetric or fluorescent reagent-based protein assay techniques have been developed. They are widely used by laboratories involved in protein research. Protein samples are added to the reagent, producing a colour change or increased fluorescence in proportion to the protein amount added. Protein concentration is determined by reference to a standard curve consisting of known concentrations of a purified reference protein (*e.g.*, lysozyme, bovine serum albumin). In ◻ Table 10.4, we summarise the principle and key features of Bradford and bicinchoninic acid (BCA) assays.

10.3 Protein Characterization

Having completed protein purification and quantification, let's move on to protein characterization, more specifically enzyme characterization. This part of the work is time consuming and it requires careful experimentation, as you are determining the enzyme kinetic parameters based upon a series of enzyme progress curves collected.

Assay	Principle	Detection mode	Detection wavelength [nm]
Bradford	Direct detection of colour change upon protein-dye (Coommasie) binding	Absorbance	595
BCA	Protein-copper chelation and secondary detection of reduced copper	Absorbance	562 (for standard BCA) and 480 (for Rapid Gold BCA)

□ **Table 10.4** Assays for protein quantification

10.3.1 Progress Curve for Enzyme-Catalysed Reaction

Reaction progress curve is a curve of absorbance (or fluorescence) value *vs* time. Using the calibration curve (or the standard curve), the absorbance (or fluorescence) value can be converted into product concentration. As such, reaction progress curve is frequently presented as a curve of product concentration *vs* time (as shown in □ Fig. 10.5).

There are two ways to derive enzyme kinetic parameters from reaction progress curve:

- Via the initial rates from a series of reaction progress curves obtained at the same enzyme concentration but at varied substrate concentrations
- Enzyme kinetic parameters directly from a reaction progress curve

10.3.1.1 Initial Rates from a Series of Progress Curves

When a low enzyme concentration is used, it is possible to capture the linear region of the reaction progress curve (*i.e.*, the initial phase; □ Fig. 10.5). The slope of the linear region will give us the initial reaction rate:

$$Initial\ reaction\ rate = slope\ of\ linear\ region = \frac{\Delta P}{\Delta t}$$

If we conduct a series of reactions at increasing substrate concentrations (say from 10 to 1000 µM), but keeping the enzyme concentration constant (0.4 nM), we will obtain the set of graphs in □ Fig. 10.6. Each line will give us one initial rate (□ Fig. 10.7). But, how do we derive the enzyme kinetic parameters (V_{max} and K_m)? Well, simple! We can do a non-linear regression (□ Fig. 10.7):

- Open Microsoft Excel and create a new workbook
- Import your kinetic data, with substrate concentration (S) in column A and initial rate in column B
- Make initial guesses of V_{max} ($G$2) and K_m ($G$3). You may ask, how do I make these guesses? If you plot 'initial rate' (y-axis) *vs* 'substrate concentration' (x-axis), V_{max} is the asymptote when substrate concentration approaches infinity (S → ∞), which is roughly 0.04 µM/s. K_m is the substrate concentration, when v = V_{max}/2. K_m is roughly 85 µM.

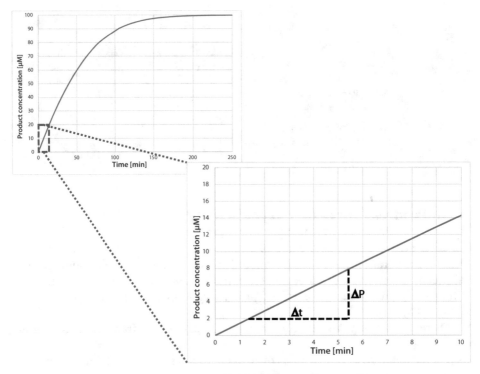

10

☐ **Fig. 10.5** The reaction progress curves of trypsin-catalysed release of *p*-nitroaniline from a trypsin pseudosubstrate N-CBZ-Gly-Pro-Arg-*p*-nitroanilide, assuming a Michaelis-Menten model. The curve is obtained using a substrate concentration of 100 μM and an enzyme concentration of 0.4 nM. During the initial phase of the reaction (see the zoomed-in portion), the product concentration increases linearly with time. The initial rate can be calculated using the slope of the linear region.

- In column C, calculate initial rate using data in column A and the equation below:

$$v = \frac{V_{max}S}{K_m + S}$$

- In column D, calculate χ^2 using the data in columns B and C, and the equation below:

$$\chi^2 = \left(Initial\ rate - Calculated\ initial\ rate \right)^2$$

- In cell $D$20, calculate the sum of χ^2
- Under the 'Tools' menu, select 'Solver'. A new pop-up window will appear.
- In the box labelled as 'Set Objective:', type $D$20 or select cell D20 in the spreadsheet
- Below this box, tick 'Min' to minimise the value in cell D20
- In the box labelled as 'By Changing Variable Cells:', type $G$2:$G$3 or select cells G2 and G3 in the spreadsheet
- Click the 'Solve' button
- The Solver will alter your initial guesses to fit the data. The best fit values for V_{max} and K_m will display in cell G2 and G3, respectively. If we plot experimentally determined data and the best curve fit in the same graph, we obtain ☐ Fig. 10.8.

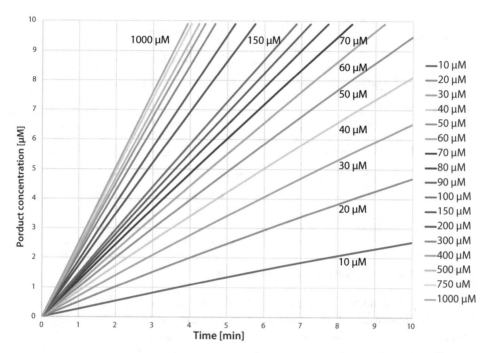

Fig. 10.6 The reaction progress curves of trypsin-catalysed release of *p*-nitroaniline from a trypsin pseudosubstrate N-CBZ-Gly-Pro-Arg-*p*-nitroanilide, assuming a Michaelis-Menten model. The curve is obtained using a substrate concentration of 10–1000 µM and an enzyme concentration of 0.4 nM.

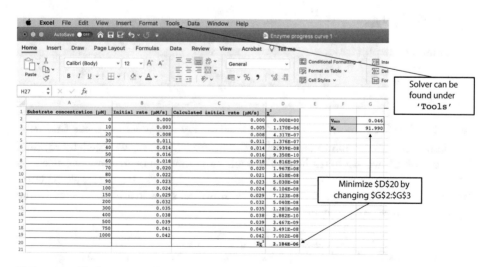

Fig. 10.7 Finding enzyme kinetic parameters from initial rates.

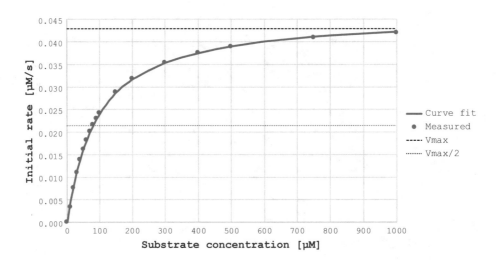

□ Fig. 10.8 Non-linear regression of enzyme kinetic data.

If you are not confident with a non-linear regression, you can of course linearize Michaelis-Menten equation, create a linear plot and perform a linear regression instead, using what you have learned in your biochemistry module:

— Lineweaver-Burk plot

$$\frac{1}{v} = \frac{K_m}{V_{\max}} \frac{1}{S} + \frac{1}{V_{\max}}$$

— Eadie-Hofstee plot

$$v = -K_m \frac{v}{S} + V_{\max}$$

— Hanes-Woolf plot

$$\frac{S}{v} = \frac{S}{V_{\max}} + \frac{K_m}{V_{\max}}$$

10.3.1.2 Kinetic Parameters Derived Directly from a Progress Curve

To obtain the kinetic parameters using the approach above, we need to conduct 18 reactions. If we do a triplicate for each substrate concentration, we need 54 ($18 \times 3 = 54$) reactions. Is it possible to deduce $V_{\max}$ and K_m from a single reaction progress curve? The answer is yes. Before we show the steps involved, let's revisit the equation you have seen in ▶ Sect. 9.2.2:

$$t = \frac{P}{k_2 E} + \frac{K_m}{k_2 E} \ln \frac{S_0}{S_0 - P} = \frac{P}{V_{\max}} + \frac{K_m}{V_{\max}} \ln \frac{S_0}{S_0 - P} = \frac{P}{V_{\max}} - \frac{K_m}{V_{\max}} \ln \frac{S_0 - P}{S_0}$$

The equation above can be simplified, if we introduce fractional substrate remained (X), which is defined as:

$$X = \frac{S}{S_0} = \frac{S_0 - P}{S_0}$$

The relationship between P and t can now be rewritten as:

$$t = \frac{S_0}{V_{max}}(1-X) - \frac{K_m}{V_{max}}\ln X = A(1-X) - B\ln X$$

$$A = \frac{S_0}{V_{max}}$$

$$B = \frac{K_m}{V_{max}}$$

Let's do another non-linear regression (◘ Fig. 10.9):
- Open Microsoft Excel and create a new workbook
- Import your reaction progress curve, with product concentration (P) in column A and time (t) in column B
- In column C, calculate X
- Make initial guesses of A ($H$2) and B ($H$3).
- In column D, calculate time using data in column C and the equation below:

$$t = A(1-X) - B\ln X$$

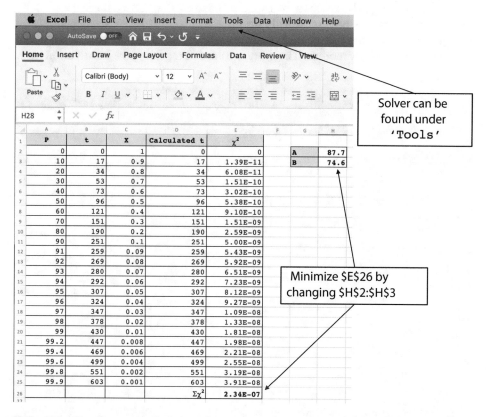

◘ **Fig. 10.9** Non-linear regression of reaction progress data in ▶ Fig. 9.5 for trypsin-catalysed release of *p*-nitroaniline from a trypsin pseudosubstrate N-CBZ-Gly-Pro-Arg-*p*-nitroanilide, assuming a Michaelis-Menten model. The curve is obtained using a substrate concentration of 100 μM and an enzyme concentration of 10 nM.

- In column E, calculate χ^2 using the data in columns B and D, and the equation below:

$$\chi^2 = \left(time - Calculated\ time\right)^2$$

- In cell \$E\$26, calculate the sum of χ^2
- Under the 'Tools' menu, select 'Solver'. A new pop-up window will appear.
- In the box labelled as 'Set Objective:', type \$E\$26 or select cell E26 in the spreadsheet
- Below this box, tick 'Min' to minimise the value in cell E26
- In the box labelled as 'By Changing Variable Cells:', type \$H\$2:\$H\$3 or select cells H2 and H3 in the spreadsheet
- Click the 'Solve' button
- The Solver will alter your initial guesses to fit the data. The best fit values for A and B will display in cell H2 and H3, respectively. With these new values, you can calculate V_{max} and K_m values.

10.3.2 Enzyme Inhibition

In the V_{max} and K_m calculations above, we are assuming that there is no enzyme inhibition. If an inhibitor is present, it will affect the V_{max} and/or K_m value, depending on the type of inhibition (◘ Table 10.5). Enzyme inhibition is a topic taught in most fundamental biochemistry module. Therefore, we are not repeating the information here.

◘ **Table 10.5** Enzyme inhibition

Type of inhibition	Details	Reaction rate	IC$_{50}$
Competitive	Competitive inhibitor competes with the substrate at the active site, and therefore increases the K_m value	$v = \dfrac{V_{max}}{1 + \dfrac{K_m}{S}\left(1 + \dfrac{I}{K_i^C}\right)}$	$\dfrac{K_i}{I}\left(1 + \dfrac{S}{K_m}\right)$
Non-competitive	Non-competitive inhibitor binds to another location on the enzyme. As such, it decreases the V_{max} value, and not the K_m value.	$v = \dfrac{V_{max}}{\left(1 + \dfrac{K_m}{S}\right)\left(1 + \dfrac{I}{K_i}\right)}$	$K_i\left(1 + \dfrac{K_m}{S}\right)$
Uncompetitive	Uncompetitive inhibition occurs when the inhibitor binds only to the enzyme-substrate complex and not the free enzyme. It decreases both the V_{max} and the K_m value.	$v = \dfrac{V_{max}}{\left(1 + \dfrac{I}{K_i^U}\right) + \dfrac{K_m}{S}}$	K_i
Mixed	The inhibitor binds to the enzyme whether or not the enzyme has already bound the substrate. Conceptually, it can be seen as a mixture of competitive inhibition and uncompetitive inhibition.	$v = \dfrac{V_{max}}{\left(1 + \dfrac{I}{K_i^U}\right) + \dfrac{K_m}{S}\left(1 + \dfrac{I}{K_i^C}\right)}$	$\dfrac{S + K_m}{\dfrac{S}{K_i'} + \dfrac{K_m}{K_i}}$

1. DNA sequencing is a direct way to determine the nature and the location of the mutation(s) in the gene encoding an improved protein variant.

2. A successful DNA sequencing reaction gives a chromatogram that is characterised by defined and evenly spaced peaks with high signal-to-noise ratio and a quality score of >40.

3. Two mutations are considered to be purely additive (or multiplicative), if the effect of the double mutation is the sum of the effects of the single mutations.

4. Epistasis describes the phenomenon in which the effect of one mutation is dependent on the presence or absence of the other mutation.

5. Capture, Intermediate Purification and Polishing (CIPP) are the three stages in a protein purification scheme.

6. Purified protein can be quantified using either its extinction coefficient or protein assays.

7. Bradford and bicinchoninic acid (BCA) assays are commonly used for protein quantification.

8. Enzyme kinetic parameters can be obtained in two ways: (a) via the initial rates from a series of reaction progress curves obtained at the same enzyme concentration but at varied substrate concentrations, and (b) directly from a reaction progress curve.

9. Excel's Solver is an excellent tool for non-linear regression.

Exercise

(a) If you clone Tfu_0901$_{37-466}$ into pGEX-4T-1 using restriction sites EcoRI and NotI (◘ Fig. 10.10), what protein purification scheme would you use?

(b) What chromatographic columns would you use?

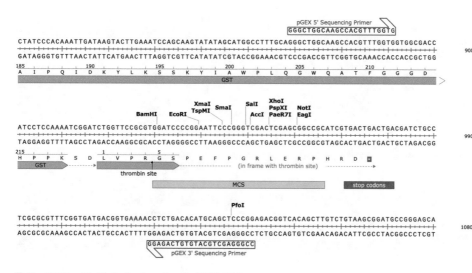

◘ **Fig. 10.10** Multiple cloning site of pGEX-4T-1.

(c) What purification buffers would you use for each purification step?

(d) What protease would you use to remove the GST tag?

(e) What are the MW and pI values, pre- and post-cleavage of GST tag?

(f) What are the molar extinction coefficients, pre- and post-cleavage of GST tag?

Further Reading

Goldring JP (2012) Protein quantification methods to determine protein concentration prior to electro-phoresis. Methods Mol Biol 869:29–35

Scopes RK. (2001). Overview of protein purification and characterization. Curr Protoc Protein Sci Chapter 1:Unit 1.1

Storz JF (2018) Compensatory mutations and epistasis for protein function. Curr Opin Struct Biol 50:18–25

Continuing from Protein Variants

Contents

© Springer Nature Switzerland AG 2020
T. S. Wong, K. L. Tee, *A Practical Guide to Protein Engineering*, Learning Materials in Biosciences,
https://doi.org/10.1007/978-3-030-56898-6_11

What You Will Learn in This Chapter

In this Chapter, we will learn to:

— perform DNA recombination using Stemmer shuffling or StEP.
— combine or dilute mutations using DNA recombination methods such as StEP.
— perform saturation mutagenesis.
— understand various codon randomization schemes.

We see protein engineering as a learning process. The more we work on a protein, the more we understand the protein. We firmly believe there is 'no end to learning'. There are a lot more that we can do, once we have found improved protein variants. It is impossible for us to cover all the possibilities within one book, instead we will discuss simple methods that can be used to further your protein engineering campaign. In this final technical chapter, we will look at two additional gene mutagenesis methods:

— DNA recombination
— Saturation mutagenesis

Using DNA recombination, you can either combine the beneficial mutations you have found or dilute the mutations should you want to study the interactions between these mutations. With saturation mutagenesis, also known as sequential permutation, it allows you to systematically replace wildtype amino acid with all 19 non-wildtype amino acids at the position identified in your directed evolution experiment.

As with all earlier chapters in this book, we will use Tfu_0901_{37-466} as an example to illustrate how these techniques can be applied.

11

11.1 DNA Recombination

DNA recombination is the exchange of DNA strands to produce new nucleotide sequence arrangements. Recombination occurs typically, although not exclusively, between homologous regions by breaking and re-joining DNA segments. It is an essential mechanism for creating genetic diversity and maintaining genome integrity as well.

In the early 1990s, Willem 'Pim' Stemmer (1957–2013) invented a protein engineering technique called the DNA shuffling. The method mimics the natural DNA recombination, and it revolutionized the engineering of therapeutic proteins, vaccines and industrial enzymes. He was recognized for his seminal work by the 2011 Charles Stark Draper Prize, which he shared with Frances H. Arnold. Draper Prize is the US's top engineering honour, often dubbed as the Nobel Prize in Engineering.

There are two stages in DNA shuffling (◘ Fig. 11.1). In stage 1, the starting pool of genes is randomly fragmented using DNase I. In the second stage, the random fragments are reassembled and amplified in a PCR. A critical step in DNA shuffling is the gene fragmentation to produce fragments of appropriate size through the control of DNA concentration, DNase I concentration and digestion time. Important to highlight, DNA shuffling can be applied to a single gene (*e.g.*, multiple gene variants identified from a random mutagenesis library) or multiple genes (*e.g.*, a set of homologous gene sequences). Sequences with DNA fragments derived from a set of parental sequences are commonly known as chimeras.

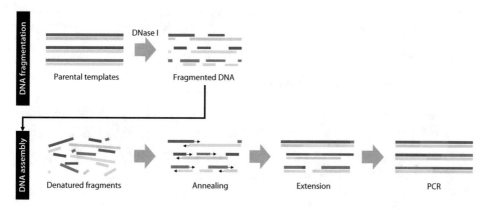

Fig. 11.1 The procedure of Stemmer shuffling.

Table 11.1 Commercial kits for DNA recombination

Kit	Supplier	Principle
JBS DNA-Shuffling Kit	Jena Bioscience	Stemmer shuffling
PickMutant™ DNA Shuffling Kit	Canvax	Stemmer shuffling

As with other gene mutagenesis methods, commercial kits for DNA recombination are available (■ Table 11.1). Perhaps not surprising, they are based on the principle of DNA shuffling. These kits come with clear manufacturer instructions, which we will not repeat here.

11.1.1 Staggered Extension Process (StEP)

Huimin Zhao and Frances H. Arnold developed a simple method for homologous DNA recombination called the Staggered Extension Process (StEP), without the need of DNA fragmentation using DNase I. StEP consists of priming the template sequences, followed by repeated cycles of denaturation and extremely abbreviated annealing/polymerase-catalysed extension. As shown in ■ Fig. 11.2, in each cycle, the growing fragments anneal to different templates based on sequence complementarity and extend further. This process is repeated until the full-length sequences are formed. Due to the constant template switching, most of the daughter DNA sequences will have a combined sequence information from different parental sequences.

To put it simply, StEP is analogous to performing a PCR. The key differences between StEP and a standard PCR for gene amplification are:

- In StEP, multiple gene sequences (*e.g.*, multiple gene variants identified from a random mutagenesis library or multiple homologous gene sequences) are used as template DNAs, instead of a single DNA template in a standard PCR.
- The thermal cycling conditions of StEP are different, which include very brief annealing/extension.

Parental templates   Priming   Staggered extension   Library

■ **Fig. 11.2**   Staggered Extension Process (StEP).

■ **Table 11.2**   Preparing a StEP mixture

Component	Volume [µL]
Water	64.5 − X
10× QIAGEN PCR buffer	10
5× Q-solution[a]	20
10 mM dNTP	2
20 µM Fwd primer[b]	1.5
20 µM Rev primer[c]	1.5
Plasmid DNA[d]	X (0.15 pmol)
Taq DNA polymerase (5 U/µL)	0.5
Total	**100**

[a]To enhance PCR efficiency of template with a high GC content
[b]Use the Fwd primer designed in ▶ Sect. 5.1.3.1
[c]Use the Rev primer designed in ▶ Sect. 5.1.3.1
[d]Plasmids harbouring gene variants identified in random mutagenesis. See ▶ Sect. 8.4.1.2

■ **Table 11.3**   Thermal cycling conditions of StEP

Temperature [°C]	Duration	# of cycle
95	5 min	1×
94	30 s	80×
55	5 s	
8	∞	–

■ Tables 11.2 and 11.3 show the PCR composition and thermo cycling conditions when performing StEP for $Tfu\_0901_{37\text{-}466}$. You may be puzzled by the missing extension step during the thermal cycling. The success of StEP hinges on abbreviated annealing/extension. As the temperature is ramping up from 55 to 94 °C, there will be a momentary strand extension when temperature passes the 72 °C mark (which is the recommended extension temperature for Taq DNA polymerase). The requirement of

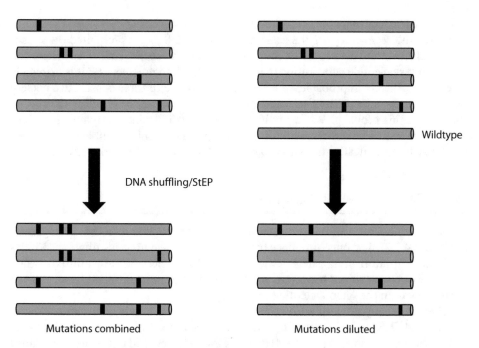

● **Fig. 11.3** Combining or diluting mutations using DNA recombination approaches.

a brief extension in each cycle would mean that the use of high processivity or fast DNA polymerases should be avoided when setting up StEP.

11.1.2 Diluting Mutations Using DNA Recombination Methods

So far, our discussion of DNA recombination has been focusing on combining mutations (● Fig. 11.3). However, more is not always good. You may recall the interactions between mutations discussed in ▶ Sect. 10.1.3. Some mutations are deleterious. When removed, the phenotypic enhancement is even greater. Experimentally, this can be easily achieved by adding a wildtype sequence into the pool of genes that are used as templates in DNA recombination, as illustrated in ● Fig. 11.3. The extent of mutational dilution can be adjusted by tuning the proportion of the wildtype sequence in the parental gene pool.

11.2 Saturation Mutagenesis

The key objective of saturation mutagenesis is to substitute a targeted amino acid with its 19 counterparts in order to study the phenotypic effect from 20 amino acids covering all physicochemical properties. Random mutagenesis (▶ Sect. 8.4.1) and saturation mutagenesis are complementary to one another. Random mutagenesis allows you to identify residues that are critical for phenotypic enhancement. However, the redundancy (or the degeneracy) of the genetic code (▶ Sect. 8.2) has prohibited the substitutions of an amino acid to its 19 counterparts. This is because statistically there can only be one nucleotide substitution in a codon in random mutagenesis; it is extremely rare, or statistically improbable, to have two consecutive or three consecutive nucleotide substitutions in random mutagenesis. Saturation mutagenesis offers the precise solution to this problem. While random mutagenesis introduces random mutations into the entire gene sequence, saturation mutagenesis can only be applied to targeted positions. It is practically impossible to saturate every single codon in a gene sequence.

11.2.1 Randomization Scheme

Experimentally, saturation mutagenesis is similar to site-directed mutagenesis (▶ Sect. 8.4.2.2). The only difference between them is the mutagenic oligonucleotides used. In saturation mutagenesis, degenerated oligonucleotides are utilised, which contain wobbles or mixed bases (*i.e.*, mixtures of two or more different bases at given positions within the oligonucleotide sequence).

   There are many randomization schemes to randomize a codon, as summarised in ◘ Fig. 11.4. The easiest option is NNN, where N represents any nucleotides (A/T/G/C). There are 64 codons ($4 \times 4 \times 4$) in NNN for 20 amino acids, and this redundancy increases significantly as the number of codons to be saturated increases (◘ Fig. 11.5). As an example, if we are to randomize two codons in a gene sequence, the number of unique protein sequences is 400 (20×20). When we apply NNN randomization scheme, the number of genes in our library is 4096 (64×64). This represents a 10.24-fold redundancy ($4096/400 = 10.24$), the value indicated in the abscissa of ◘ Fig. 11.5. There are another two drawbacks of using NNN randomization scheme:
- There are 3 stop codons in NNN, resulting in prematurely truncated protein sequences.
- There is a significant concentration difference between individual protein sequences. For example, serine is encoded by 6 codons, and methionine is encoded by only 1 codon in NNN randomization scheme (◘ Fig. 11.4). The concentration difference between the protein sequence with a serine residue at the targeted position and the protein sequence with a methionine residue is 6 folds. This concentration difference increases exponentially, as the number of codons to be saturated increases (◘ Fig. 11.6). This concentration difference is reflected in the bias indicator in ◘ Fig. 11.4.

To overcome the three problems mentioned above (redundancy, stop codons and concentration difference), one can use NNB, NNK or NNS randomization schemes (◘ Fig. 11.4). With these schemes, the extents of all three problems are drastically reduced. B, K and S refers to T/G/C, G/T and G/C mixtures, respectively. NDT is another interesting randomization scheme to consider. Although not all 20 amino acids are covered in NDT, there are representatives from each amino acid categories (aliphatic, aromatic, neutral and charged) in the library (also known as a smart library).

Property	Amino acid	NNN	NNB	NNK	NNS	NDT	NDT+VMA+ATG+TGG	NDT+VHG+TGG	MAX
Number of primer pairs		1	1	1	1	1	4	3	1
Ratio		1	1	1	1	1	12:6:1:1	12:9:1	1
Aliphatic	G	4	3	2	2	1	1	1	1
	A	4	3	2	2	0	1	1	1
	V	4	3	2	2	1	1	2	1
	L	6	4	3	3	1	1	2	1
	I	3	2	1	1	1	1	1	1
Aromatic	F	2	2	1	1	1	1	1	1
	Y	2	2	1	1	1	1	1	1
	W	1	1	1	1	0	1	1	1
Neutral	C	2	2	1	1	1	1	1	1
	M	1	1	1	1	0	1	1	1
	P	4	3	2	2	0	1	1	1
	S	6	5	3	3	1	1	1	1
	T	4	3	2	2	0	1	1	1
	N	2	2	1	1	1	1	1	1
	Q	2	1	1	1	0	1	1	1
Charged	D	2	2	1	1	1	1	1	1
	E	2	1	1	1	0	1	1	1
	H	2	2	1	1	1	1	1	1
	K	2	1	1	1	0	1	1	1
	R	6	4	3	3	1	1	1	1
Stop codon	*	3	1	1	1	1	0	0	0
Number of codons		64	48	32	32	12	20	22	20
Bias indicator (L/S/R:M/W)		6	4 or 5	3	3	No M/W	1	1 or 2	1

▫ Fig. 11.4 Various randomization schemes for saturation mutagenesis.

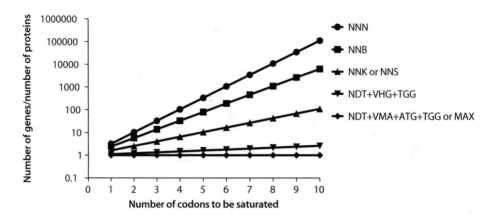

Fig. 11.5 The effect of increasing the number of codons to be saturated on the gene library redundancy. The ratios on the abscissa are calculated by dividing the number of gene permutations in a randomised library by the number of encoded proteins (20^N), where N is the number of codons to be saturated.

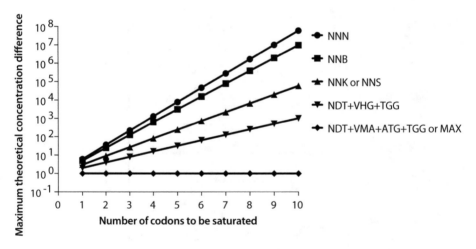

Fig. 11.6 The effect of increasing the number of codons to be saturated on the maximum theoretical concentration difference.

Instead of using one pair of mutagenic oligos to randomize a codon (QuikChange approach requires one pair of mutagenic oligos per mutated site), we can also use multiple pairs of mutagenic oligos. For instance, if we mix NDT ($4 \times 3 \times 1 = 12$ codons), VMA ($3 \times 2 \times 1 = 6$ codons), ATG (1 codon) and TGG (1 codons) in a ratio of 12:6:1:1, all 20 amino acids are covered with no redundancy, no stop codon and no concentration difference. Mixing NDT ($4 \times 3 \times 1 = 12$ codons), VHG ($3 \times 3 \times 1 = 9$ codons) and TGG in 12:9:1 ratio will give an almost similar outcome, but with one pair of mutagenic oligos less. MAX randomization scheme employs specialised trinucleotide phosphoramidites during the chemical synthesis of oligo-nucleotides, therefore it does not require mixing of mutagenic oligos.

11.2.2 **Designing a Saturation Mutagenesis Experiment**

Since a saturation mutagenesis experiment is similar to a site-directed mutagenesis. Our recommendation is to design your experiment using OneClick (▶ http://www. oneclick-mutagenesis.com/, ▶ Sect. 8.4.2.3). OneClick has incorporated NNK and NDT randomization schemes into its algorithm.

Take-Home Messages

1. Willem 'Pim' Stemmer invented the protein engineering technique called DNA shuffling in the early 1990s.
2. Staggered Extension Process (StEP) is a simple method for homologous DNA recombination.
3. DNA recombination can be used to either combine or dilute mutations found in random mutagenesis.
4. The key objective of a saturation mutagenesis is to substitute a targeted amino acid with its 19 counterparts in order to study the phenotypic effect from 20 amino acids covering all physicochemical properties.
5. There are many randomization schemes to randomize a codon. They differ in the number of codons, the number of stop codons, the number of amino acids encoded, the level of redundancy, the concentration difference between protein members, and the oligonucleotide synthesis.
6. OneClick is a tool for designing focused mutagenesis experiments, including saturation mutagenesis.

Exercise

(a) How many unique protein sequences are there in a library, when four codons are saturated?
(b) If we apply NNK randomization scheme for the task in (a), how many folds of redundancy are we expecting?
(c) What is the maximum theoretical concentration difference if we apply NNK randomization scheme for the task in (a)?
(d) Repeat tasks (b) and (c) using NNB randomization scheme.

Further Reading

Acevedo-Rocha CG, Reetz MT, Nov Y (2015) Economical analysis of saturation mutagenesis experiments. Sci Rep 5:10654

Stemmer WP (1994) Rapid evolution of a protein *in vitro* by DNA shuffling. Nature 370(6488):389–391

Zhao H, Giver L, Shao Z, Affholter JA, Arnold FH (1998) Molecular evolution by staggered extension process (StEP) *in vitro* recombination. Nat Biotechnol 16(3):258–261

Employability

Contents

© Springer Nature Switzerland AG 2020
T. S. Wong, K. L. Tee, *A Practical Guide to Protein Engineering*, Learning Materials in Biosciences,
https://doi.org/10.1007/978-3-030-56898-6_12

What You Will Learn in This Chapter

In this chapter, we will learn to:

- understand the technical expertise of a protein engineer.
- understand the job prospects of a protein engineer.
- understand the skill set sought after by job recruiters.
- make good use of career resources available to improve your job application and plan your career.
- apply for a job through job advertising sites.

Protein engineers are technically versatile, as we have a diverse and highly sought-after technical skill set, outlined in ◘ Fig. 12.1. But, being technically competent is not sufficient to secure a good job! That's the reason why we have decided to include one more chapter in this book to discuss employability.

Employability is defined as 'a set of achievements – skills, understandings and personal attributes – that makes graduates more likely to gain employment and be successful in their chosen occupations, which benefits themselves, the workforce, the community and the economy'. For many higher education institutions, employability is now embedded in the curriculum design and delivery. Students are provided with more opportunities to engage actively in work experience. Employers and alumni are more regularly involved in universities' activities, participating as guest speakers, providing placement opportunities, informing on curriculum and assessing students. There is a greater emphasis on teaching students using real world or applied examples and applying teaching methods that help students develop their transferable skills, such as problem solving. Enterprise & entrepreneurship and industrial management are examples of optional modules offered to

12

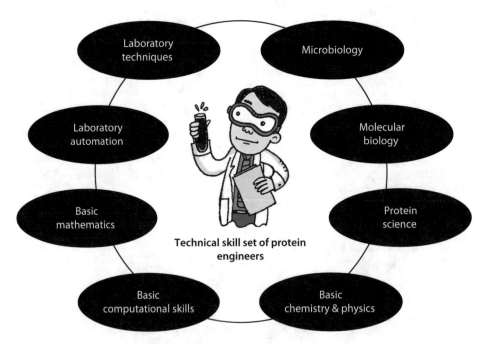

◘ **Fig. 12.1**   Technical skill set of protein engineers.

students to diversify their knowledge. The role of a lecturer has also expanded to include providing career advice, motivating and supporting student's employability development.

12.1 Sectors Requiring Protein Engineering Expertise

Although protein engineering is a specialised subject, there are many sectors that require protein engineering expertise:

- Enzyme companies (*e.g.*, Novozymes, New England Biolabs, Amano, and AB Enzymes *etc*)
- Chemical companies (*e.g.*, BASF, DSM, DuPont, Evonik, Mitsubishi Chemical, and Mitsui Chemicals *etc*)
- Pharmaceutical and diagnostic companies (*e.g.*, Roche, GSK, AstraZeneca, Pfizer, Amgen, Merck, Novartis, Novo Nordisk and Lonza *etc*)
- Food and feed companies (*e.g.*, Cargill)
- Healthcare and personal care companies (*e.g.*, Johnson & Johnson)
- Industrial biotechnology companies (*e.g.*, Deinove and NatureWorks *etc*)
- Small and medium enterprises (SMEs)
- Start-up companies
- Academia and research institutions

The list above is, by no means, exhaustive. The key messages we want to bring across are (a) there are plenty of opportunities out there, and (b) be flexible! Keep an open mind, apply the skill set you have and be ready to learn on the job. Also useful to note, a postgraduate study (Master or PhD) is a career option. Students should not overlook this option, especially if they have a passion for research.

12.2 What Are the Skills Employers Are Looking for?

Employers are placing a lot of emphasis on finding candidates with the right skills and competencies for their organisations, and those who can integrate into existing teams. The top ten skills recruiters want are:

1. Commercial awareness or business acumen: This is all about knowing what a business is, how a business functions, and what makes a company tick.
2. Communication: This covers all forms of communication, verbal, written as well as listening.
3. Teamwork: This refers to being a team player and demonstrating the ability to manage and delegate to others.
4. Negotiation and persuasion: This is about setting your target and developing a pathway to achieve it, but also being able to understand where the other person is coming from so that you can both get what you want or need and feel positive about it.
5. Problem solving: This is the ability to take a logical and analytical approach to solving problems and resolving issues or conflicts.
6. Leadership: This includes managing people and motivating others to work for you.

7. Organization: This concerns prioritizing tasks, working efficiently and meeting deadlines.
8. Perseverance and motivation: All employers want positive thinkers, who can get through the challenges or crisis they face.
9. Ability to work under pressure: This means remaining calm in a crisis or during a stressful period.
10. Confidence: In the workplace you need to be confident in yourself but not arrogant. You are also expected to have confidence in your colleagues, people working for you and the company you work for.

These skills take time to develop. No one is born with all these skills. It is therefore important to self-assess and self-reflect to identify skill gaps and weaknesses so that you can work on them.

12.3 How to Improve Your Curriculum Vitae (CV)?

A curriculum vitae (CV) is a summary of your achievements and experiences. Most of the time, this is the very first document your employer or recruiter will look at. Therefore, you need to make sure it is as good as it can be. It takes time to build a good CV, as you need to accumulate work experience, develop skill set and demonstrate your competence through award/honour/recognition.

Below you'll find some good tips for improving your CV:
- Do a summer placement or a year-long placement to acquire work experience and develop skills. If you fail to secure a placement, marshal your resources and ask your parents, friends and relatives if they need an extra pair of hands.
- Participate in extracurricular activities.
- Join student competitions [*e.g.*, International Genetically Engineered Machine (iGEM)].
- Learn something by yourself (*e.g.*, a language, a programming language or Arduino *etc*).
- Do a summer research project (*e.g.*, Summer Undergraduate Research Fellowship) with your lecturers.
- Volunteer in your local community.

12.4 Career Planning

It is important to consider career planning when you begin higher education. We strongly advise against delaying it till you graduate or are close to completing your study. Start planning as soon as you can by:
- Attending career fairs, talking to employers, and understanding their businesses
- Networking with alumni and industrial guests
- Discussing with your personal tutor
- Seeking advice from career services
- Speaking to your parents
- Reading more widely

12.5 Where to Look for Career Resources?

There are plenty of useful resources for career development. Below we have cherry picked a few resources which we trust you will find useful:

— TARGETjobs (▶ https://targetjobs.co.uk/)
— Prospects (▶ https://www.prospects.ac.uk/)

These resources offer excellent tips on writing CV and cover letter, developing online profiles (*e.g.*, LinkedIn), preparing for all formats of interview (face-to-face, phone, video call) and types (technical, informational, open-ended, situational, behavioral *etc*), assessment centres and psychometric tests.

12.6 Where to Look for a Job?

When you are ready for job hunting, opportunities can be identified from the resources listed below:

— ▶ https://www.jobs.ac.uk/
— ▶ https://www.findaphd.com/
— ▶ https://www.nature.com/naturecareers
— ▶ https://jobs.newscientist.com/en-gb/
— ▶ https://www.jobsinscience.com/
— ▶ https://www.linkedin.com/

To end this book, we wish you the best of luck with your protein engineering career!

Take-Home Messages

1. Protein engineers are technically versatile.
2. Protein engineering expertise is highly sought after by diverse sectors.
3. The top ten skills recruiters are looking for are commercial awareness, communication, teamwork, negotiation & persuasion, problem solving, leadership, organization, perseverance & motivation, ability to work under pressure, and confidence.
4. It takes time to build a strong curriculum vitae (CV).
5. Career planning should begin as soon as possible.
6. There are a host of online resources offering excellent tips on writing CV and cover letter, developing online profiles, preparing for all formats of interview, assessment centres and psychometric tests.
7. ▶ https://www.findaphd.com/ is a good resource for finding PhD opportunities.
8. ▶ https://www.linkedin.com/ is developing itself into a good channel for job advertising and job hunting.

Exercise

(a) Conduct a self-assessment and identify your strengths and skill gaps.
(b) If you are to prepare a CV for yourself, what information would you include?
(c) Do you think your CV in (b) is strong? If not, how could you improve your CV?
(d) Who can provide personal reference for you, when you apply for a job? Why are these the right people to provide a reference?
(e) What makes you stand out from the rest of the applicants, when you apply for a job?
(f) Where do see yourself in 10-year time from the first job your apply? What is your career goal?

Further Reading

Fahnert B (2015) On your marks, get set, go!-lessons from the UK in enhancing employability of graduates and postgraduates. FEMS Microbiol Lett 362(19)